S0-AWE-984

F. W. Lipps
1986

Applied Mathematical Sciences
Volume 61

Editors
F. John   J.E. Marsden   L. Sirovich

Advisors
M. Ghil   J.K. Hale   J. Keller
K. Kirchgässner   B.J. Matkowsky
J.T. Stuart   A. Weinstein

Applied Mathematical Sciences

(continued on inside back cover)

Sattinger, David H.

D.H. Sattinger
O.L. Weaver

Lie Groups and Algebras with Applications to Physics, Geometry, and Mechanics

With 25 Illustrations

Springer-Verlag
New York Berlin Heidelberg Tokyo

QA
1
.A647
v.61

D.H. Sattinger
School of Mathematics
University of Minnesota
Minneapolis, Minnesota 55455
U.S.A.

O.L. Weaver
Department of Physics
Kansas State University
Manhattan, Kansas 66506
U.S.A.

Editors

F. John
Courant Institute of
  Mathematical Sciences
New York University
New York, NY 10012
U.S.A.

J.E. Marsden
Department of
  Mathematics
University of California
Berkeley, CA 94720
U.S.A.

L. Sirovich
Division of Applied
  Mathematics
Brown University
Providence, RI 02912
U.S.A.

AMS Subject Classifications: 22-01, 22E46, 17B05, 17B10, 34A34, 34C35

Library of Congress Cataloging in Publication Data
Sattinger, David H.
  Lie groups and algebras with applications to physics,
geometry, and mechanics.
  (Applied mathematical sciences; v. 61)
  Bibliography: p.
  Includes index.
  1. Lie groups. 2. Lie algebras. I. Weaver, O.L.
II. Title. III. Series: Applied mathematical sciences
(Springer-Verlag New York Inc.); v. 61.
QA1.A647   vol. 61   [QA387]   510 s   [512'.55]   85-27856

© 1986 by Springer-Verlag New York Inc.
All rights reserved. No part of this book may be translated or reproduced in any form
without written permission from Springer-Verlag, 175 Fifth Avenue, New York, New
York 10010, U.S.A.

Typeset by Asco Trade Typesetting Ltd., Hong Kong.
Printed and bound by R.R. Donnelley & Sons, Harrisonburg, Virginia.
Printed in the United States of America.

9 8 7 6 5 4 3 2 1

ISBN 0-387-96240-9 Springer-Verlag New York Berlin Heidelberg Tokyo
ISBN 3-540-96240-9 Springer-Verlag Berlin Heidelberg New York Tokyo

Preface

This book is intended as an introductory text on the subject of Lie groups and algebras and their role in various fields of mathematics and physics. It is written by and for researchers who are primarily analysts or physicists, not algebraists or geometers. Not that we have eschewed the algebraic and geometric developments. But we wanted to present them in a concrete way and to show how the subject interacted with physics, geometry, and mechanics. These interactions are, of course, manifold; we have discussed many of them here—in particular, Riemannian geometry, elementary particle physics, symmetries of differential equations, completely integrable Hamiltonian systems, and spontaneous symmetry breaking.

Much of the material we have treated is standard and widely available; but we have tried to steer a course between the descriptive approach such as found in Gilmore and Wybourne, and the abstract mathematical approach of Helgason or Jacobson. Gilmore and Wybourne address themselves to the physics community whereas Helgason and Jacobson address themselves to the mathematical community. This book is an attempt to synthesize the two points of view and address both audiences simultaneously. We wanted to present the subject in a way which is at once intuitive, geometric, applications-oriented, mathematically rigorous, and accessible to students and researchers without an extensive background in physics, algebra, or geometry.

Our operating assumption is that our reader has a good founding in linear algebra and some familiarity with the basic ideas of group theory which are traditionally taught at the undergraduate level. For example, we assume he/she knows what a normal subgroup is and how a quotient group is formed. We also assume the reader is familiar with such elementary topological notions as continuity, compactness, and simple connectivity. We discuss in some detail the tensor calculus on manifolds, but we have avoided the technicalities of differentiable structures on manifolds. Some knowledge of differential geometry and tensor analysis would be helpful for a complete understanding of some of the specialized topics in geometry and mechanics.

Whether this book is a public success or not, it has certainly been a private one. In bridging the "jargon gap" between mathematician and physicist we have personally gained a deeper understanding of the material. One outcome of this process, for example, is our presentation of the material in Chapter 5 traditionally known as "tensor calculus on manifolds." We have presented that material here in the physicist's language of frame invariance, not the mathematician's notation of pull-backs (φ^* and φ_* and all that).

We have tried to introduce and use the language of differential forms in a way that will induce the physicist to learn the approach. The calculus of differential forms has had many promoters—the chief one being Cartan himself, and that alone should be ample recommendation. We discuss in some detail the Maurer–Cartan forms on a Lie group and their applications. For example, in §26 (Geometry "à la Cartan") we explain Cartan's derivation of the structure equations of Riemannian manifolds from the Maurer–Cartan equations of their isometry groups. This leads, for example, to the structure equations of surfaces of constant curvature, as well as to the structure equations of a surface embedded in R^3. It touches upon Cartan's brilliant theory of symmetric spaces, a topic which we have treated all too briefly.

The use of differential forms has not yet achieved widespread acceptance by the physics or applied mathematics community, but appears to be gaining ground. Differential forms, being dual to vector fields, are necessarily more abstract and less easily grasped as fundamental intuitive objects. Yet the exterior differential calculus is an excellent form of bookkeeping that takes into account orientation, provides the correct language for multidimensional integration, and is "frame independent"—that is, invariant under arbitrary coordinate transformations. It is perhaps worthwhile comparing the situation with the acceptance of Gibbs' vector notation, which (according to Struik in his text *Differential Geometry*) "after years of competition with other notations seems to have won the day...." This in 1950! It appears that differential forms are still competing but have not yet won the day.

We have benefited from many discussions with our colleagues, in particular Leon Green, Robert Ellis, Paul Garrett, and Sid Webster at the University of Minnesota. We also thank the students who took the course at Minnesota, who asked so many good questions, and made so many valuable comments: Russell Brown, Susan Fischel, David Gregg, Gerald Warnecke, and Victor Zurkowski.

Minneapolis, Minnesota   D.H. SATTINGER
Manhattan, Kansas   O.L. WEAVER
August 1985

Contents

LIE GROUPS AND ALGEBRAS

CHAPTER 1

Lie Groups

1. Continuous Groups; Covering Groups

Sophus Lie (1842–1899) and Felix Klein (1849–1925) were students together in Berlin in 1869–70 when they conceived the notion of studying mathematical systems from the perspective of the transformation groups which left these systems invariant. Thus Klein, in his famous *Erlanger* program, pursued the role of finite groups in the studies of regular bodies and the theory of algebraic equations, while Lie developed his notion of continuous transformation groups and their role in the theory of differential equations. Lie's work was a *tour de force* of the 19th century, and today the theory of continuous groups is a fundamental tool in such diverse areas as analysis, differential geometry, number theory, differential equations, atomic structure, and high energy physics. This book is devoted to a careful exposition of the mathematical foundations of Lie groups and algebras and a sampling of their applications in differential equations, applied mathematics, and physics.

In this first chapter you will be introduced to a variety of important Lie groups, together with some of their properties.

A topological group is a group which is also a topological space (so that ideas such as continuity, connectedness and compactness apply) in which the group operations are continuous. A Lie group is a topological group which is also an analytic manifold on which the group operations are analytic. We will make this notion more precise later on.

The simplest example of a Lie group is the real line $\mathbb{R}^1$ with ordinary addition as the group operation. Similarly, $\mathbb{R}^n$ with the usual vector addition is a commutative (abelian) Lie group. Continuous matrix groups, or more generally, continuous groups of linear transformations of a vector space, are called *linear* Lie groups. For example, the set of all non-singular $n \times n$ matrices forms the group known as $GL(n, \mathbb{R})$ or $GL(n, \mathbb{C})$ depending on whether the entries are real or complex. The subset of all $n \times n$ matrices with determinant 1 forms a group called the *unimodular* or special linear group, which is denoted by $SL(n, \mathbb{R})$ or $SL(n, \mathbb{C})$. The orthogonal group, $O(n)$, is the group of $n \times n$ of matrices that satisfy $AA^t = 1$. These are a few examples of the so-called "classical" groups; we shall give a complete list at the end of this chapter.

The matrices

$$L_a = \begin{pmatrix} 1 & a \\ 0 & 1 \end{pmatrix}, \qquad a \in \mathbb{R}$$

form a linear group; and moreover,

$$L_a L_b = L_{a+b}.$$

This group is thus *isomorphic* to $\mathbb{R}^1$ and forms a *representation* of $\mathbb{R}^1$ by 2×2 matrices. In general, a *representation* of a group $\mathfrak{G}$ on a vector space V is a homomorphism from $\mathfrak{G}$ into the invertible linear transformations of V. That is $a \to t_a$ in such a way that

$$a \circ b \to t_a t_b.$$

These representations need not always be matrix representations. For example we may represent $\mathbb{R}^1$ on the infinite dimensional vector space $C^\infty(\mathbb{R})$ (the infinitely differentiable functions on the line) by

$$(T_a f)(x) = f(x + a).$$

The idea of representation helps to clarify the subtle but sometimes important distinction between an abstract group and a variety of its realizations. Thus $\mathbb{R}^1$, the set of matrices L_a, the translations T_a, and the geometric translations along $\mathbb{R}^1$ itself are all distinct but isomorphic representations of the same abstract group.

An example of a non-abelian group is the group of upper triangular matrices

$$\begin{pmatrix} 1 & a & b \\ 0 & 1 & c \\ 0 & 0 & 1 \end{pmatrix} \qquad a, b, c \in \mathbb{R}.$$

This is isomorphic to the Heisenberg group which plays a fundamental role in quantum mechanics (see Chapter 4).

The matrices

$$R(\theta) = \begin{pmatrix} \cos\theta & -\sin\theta \\ \sin\theta & \cos\theta \end{pmatrix} \qquad \theta \in \mathbb{R}$$

form a group, and $R(\theta)R(\gamma) = R(\theta + \gamma)$. These are the rotations of the plane, the group $SO(2)$. This group may also be realized as S^1, the unit circle in the complex plane with multiplication as the group operation.

The groups $\mathbb{R}^1$ and S^1 are intimately related, for the mapping $\rho\colon \mathbb{R}^1 \to S^1$ defined by $\rho(x) = e^{2\pi i x}$ is a continuous homomorphism from $\mathbb{R}^1$ onto S^1. The transformation is locally one-to-one; (see Fig. 1.1) but globally it is infinite-to-one, for all points $x + n$ (with n an integer) map onto the same point $e^{2\pi i x}$ in S^1. The kernel of ρ is Z, the discrete group of integers, and S^1 is isomorphic with the quotient group $\mathbb{R}^1/Z$. We say that each z *lifts* to an infinity of points in $\mathbb{R}^1$ (see Fig. 1.1). Moreover, the unclosed path $0 \le t \le 1$ in $\mathbb{R}^1$ is mapped onto the closed path $z(t) = e^{2\pi i t}$ in S^1. The latter cannot be shrunk to a point while remaining in S^1. Yet any closed path in $\mathbb{R}^1$ can be shrunk down to a single point. We say that $\mathbb{R}^1$ is simply connected while S^1 is not.

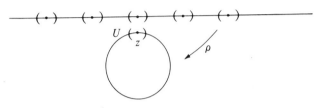

Figure 1.1. The neighborhood U of $z \in S^1$ lifts to an infinite collection of disjoint intervals in $\mathbb{R}^1$, each of which is mapped 1–1 onto U by ρ. We say ρ is locally one-to-one.

In this example, $\mathbb{R}^1$ is the *universal covering group* of S^1. The universal covering group of a connected topological group G is a simply connected topological group $\hat{G}$ together with a continuous homomorphism ρ from $\hat{G}$ onto G which is locally one-to-one. Such a universal covering group exists for every connected topological group, and in particular for every Lie group, and is unique up to isomorphism. Moreover $\hat{G}/\text{Ker }\rho \simeq G$.

Universal covering groups, their representations, and their homomorphisms into non-simply connected groups, are fundamental to our understanding of symmetry in quantum mechanics. (See Section 1 of Wigner's 1939 paper for a discussion of this point, Wigner [1].) For example, particles with half-integer spin—such as electrons, quarks, ^{3}He nuclei—transform according to *spinor representations* of the rotation and Lorentz groups. These are double valued representations of the geometrical symmetry groups, but are single-valued representations of the covering groups.

In this chapter we shall construct the universal covering groups and the covering homomorphisms for the rotation and Lorentz groups. Readers interested in a more careful treatment of covering groups can profitably consult Pontryagin, Chevalley, or Singer and Thorpe.

EXERCISES

1. Construct the covering homomorphism and the covering groups of S^1 having the form R^1/pZ where pZ is the set of integral multiples of the integer p. $\mathbb{R}^1$ is the universal covering group of these groups. Construct a five valued representation of S^1.

2. What is the universal covering group of the torus, $T^n = S^1 \times S^1 \times \cdots \times S^1$, the group whose elements are $(z_1, z_2, \ldots, z_n)$ with $|z_i| = 1$?

3. A matrix U is called unitary if $UU^* = I$. Show that the $n \times n$ unitary matrices form a group, called $U(n)$. Show that the unitary matrices of determinant 1 form a subgroup, called $SU(n)$. Show that $SU(2)$ consists of all matrices of the form

$$\begin{pmatrix} \alpha & \beta \\ -\bar{\beta} & \bar{\alpha} \end{pmatrix} \qquad |\alpha|^2 + |\beta|^2 = 1.$$

2. The Rotation Group in R^3

The group of rotations of three dimensional Euclidean space is called $SO(3)$. Every rotation $R \in SO(3)$ can be parameterized by an axis of rotation $\hat{n}$ and the angle θ of rotation about this axis: $R = (\hat{n}, \theta)$. The axis requires two angles (α, β) for its specification, so three parameters are needed to specify a general rotation: $SO(3)$ is a three parameter group. A useful visualization of the elements of $SO(3)$ is to picture a solid ball of radius π. A point P inside the ball a distance θ from the origin, 0, represents the counterclockwise rotation about the axis $\overrightarrow{OP}$ by an angle θ. Since the parameters range over a compact set, $SO(3)$ is a *compact group*; the origin 0 corresponds to the identity transformation.

A moment's reflection shows that we really do only require a ball of radius π, not 2π, but that antipodal points on the surface represent the same rotation. This enables us to show that $SO(3)$ is not a simply connected space—see path #2 in Fig. 1.2: it is not contractible to a point. We will find the simply connected covering group of $SO(3)$ later in this chapter.

Another realization of $SO(3)$ follows from the observation that rotations are linear transformations of $\mathbb{R}^3$ that preserve the inner product

$$(a, b) = \sum_{i=1}^{3} a^i b^i.$$

If R is the matrix of the linear transformation then

$$(Ra, Rb) = (a, b),$$

so

$$R^t R = 1.$$

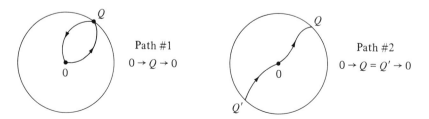

Figure 1.2

This relation shows that R is non-singular, so

$$R^t = R^{-1}.$$

These are the *orthogonal* matrices. We have actually gotten a bit more than $SO(3)$ this way, for $R = -1$ is orthogonal but is an *inversion* rather than a rotation: it reverses orientation of space. The group of orthogonal 3×3 matrices is denoted by $O(3)$, while the rotations are $SO(3)$, the special orthogonal matrices with determinant $+1$. The group $O(3)$ is not connected but is the union of the sets $\{R \in SO(3)\}$ and $\{-R, R \in SO(3)\}$. The identity, 1, is in $SO(3)$, and since $SO(3)$ is connected (although not simply connected) it is called the *connected component of the identity* in $O(3)$.

For later reference we display the orthogonal matrices corresponding to counter clockwise rotations of $\mathbb{R}^3$ about the coordinate axes:

$$R_1(\alpha) = \begin{pmatrix} 1 & 0 & 0 \\ 0 & \cos\alpha & -\sin\alpha \\ 0 & \sin\alpha & \cos\alpha \end{pmatrix},$$

$$R_2(\beta) = \begin{pmatrix} \cos\beta & 0 & \sin\beta \\ 0 & 1 & 0 \\ -\sin\beta & 0 & \cos\beta \end{pmatrix}, \tag{1.1}$$

$$R_3(\gamma) = \begin{pmatrix} \cos\gamma & -\sin\gamma & 0 \\ \sin\gamma & \cos\gamma & 0 \\ 0 & 0 & 1 \end{pmatrix}.$$

EXERCISES ON $O(3)$

1. Show that $A \to \det A$ is a homomorphism of $O(3)$ into the group $Z_2 = \{+1, -1\}$ with multiplication as the group operation. What is the kernel of this homomorphism? What are the cosets of $O(3)$ modulo that kernel? Show that $O(3)$ is not connected.

2. Given $A \in SO(3)$ show that 1 is always an eigenvalue. What is the geometrical significance of the eigenvector?

3. Show that $SO(3)$ is "doubly connected" in the sense that path #2 (Fig. 1.2) traversed twice can be shrunk to a point. Begin with the sequence of deformations shown below.

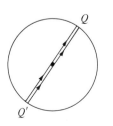 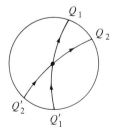 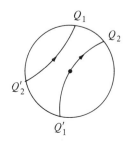

4. Projective 3-space, $P_3(\mathbb{R})$, is the set of lines through the origin in $\mathbb{R}^4$. Open sets are open cones of lines. Show that $P_3(\mathbb{R})$ is homeomorphic to the three sphere S^3 in $\mathbb{R}^4$ with antipodal points identified. Then show that this sphere is homeomorphic to the solid ball in R^3 with antipodal points on the surface identified, that is, to $SO(3)$.

3. The Möbius Group

The linear fractional transformations, or Möbius transformations, M, form the group of conformal transformations which map the extended complex plane one-to-one onto itself. An element of this group is a transformation $m: C \to C$ with

$$m(z) = \frac{az + b}{cz + d}, \qquad ad - bc \neq 0.$$

The condition $ad - bc \neq 0$ assures that the transformation is invertible. There is a homomorphism ρ from $GL(2, C)$ into M given by

$$\rho: \begin{pmatrix} a & b \\ c & d \end{pmatrix} \to m(z) = \frac{az + b}{cz + d}.$$

The reader may check that ρ indeed is a homomorphism. For all λ the matrices $\lambda \begin{pmatrix} a & b \\ c & d \end{pmatrix}$ in $GL(2, C)$ go into the *same* Möbius transformation. Since $\det\left[\lambda \begin{pmatrix} a & b \\ c & d \end{pmatrix} \right] = \lambda^2 \det \begin{pmatrix} a & b \\ c & d \end{pmatrix}$ we may always chose λ in two ways so that the determinant of $\lambda \begin{pmatrix} a & b \\ c & d \end{pmatrix}$ is $+1$. Thus each Möbius transformation is covered by two matrices of determinant one, i.e. by two elements of $SL(2, C)$.

Our homomorphism shows that M cannot be simply connected, for a path from $\begin{pmatrix} 1 & 0 \\ 0 & 1 \end{pmatrix}$ to $\begin{pmatrix} -1 & 0 \\ 0 & -1 \end{pmatrix}$ in $SL(2, C)$ is mapped by ρ onto a closed path in M which cannot be shrunk to a point. $SL(2, C)$ is a covering group of M, and $SL(2, C)$ is simply connected (see exercise 6, p. 18), so $SL(2, C)$ is the universal covering group of M.

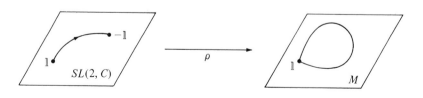

The homomorphism from $SL(2, C)$ to the Möbius transformations was obtained by Klein and Cayley in the following way. Let u and v be complex numbers and suppose

$$u' = au + bv$$
$$v' = cu + dv.$$

Putting $z = u/v$ and $w = u'/v'$ we obtain

$$w = \frac{az + b}{cz + d}.$$

The two component vectors $\begin{pmatrix} u \\ v \end{pmatrix}$ have come to be called *spinors* in physics and are used to describe particles of spin $1/2$.

The set of Möbius transformations with a, b, c, d *real* form a subgroup of Möbius transformations which preserve the upper half plane: if

$$w = \frac{az + b}{cz + d}, \qquad a, b, c, d \in \mathbb{R},$$

then $\operatorname{Im} w > 0$ whenever $\operatorname{Im} z > 0$. The associated matrices $\begin{pmatrix} a & b \\ c & d \end{pmatrix}$ form the subgroup $SL(2, R)$ of $SL(2, C)$.

The upper half plane with the metric

$$ds^2 = \frac{dx^2 + dy^2}{y^2} = \frac{dz \, d\bar{z}}{(\operatorname{Im} z)^2}$$

is known as the Poincaré half plane. It is a two dimensional Riemannian manifold with constant Gaussian curvature $K = -1$ (see Singer and Thorpe, Chapter 7). The Möbius transformations we discussed above, i.e. those which come from $SL(2, R)$, are in fact isometries of this manifold. That is, they preserve the metric tensor: if

$$w = \frac{az + b}{cz + d} \qquad a, b, c, d \in \mathbb{R}, \qquad ad - bc = 1$$

then

$$\frac{dw \, d\bar{w}}{(\operatorname{Im} w)^2} = \frac{dz \, d\bar{z}}{(\operatorname{Im} z)^2}.$$

EXERCISES

1. Prove that $SL(2, R)$ preserves the upper half plane, and the metric given above.

2. Which $SL(2, R)$ transformations preserve the norm $|u|^2 + |v|^2$ of a spinor $\begin{pmatrix} u \\ v \end{pmatrix}$?

3. Why isn't $GL(2, C)$ the universal covering group of M?

4. Show that the Möbius transformations

$$\varphi_t(z) = \frac{z + \tan t}{1 - z(\tan t)}$$

form a one-parameter subgroup, i.e. $\varphi_{t+s} = \varphi_t \circ \varphi_s$. Let $w(t) = \varphi_t(z)$ and find a differential equation for w.

4. The Covering Group of $SO(3)$

We now consider the subgroup of M that corresponds to rotations of the Riemann sphere. For this discussion the stereographic projection of $\mathbb{C}$ is useful (see Fig. 1.3).

A rotation of the Riemann sphere is a rotation in $\mathbb{R}^3$, the $\xi - \eta - \zeta$ space. Each rotation of the Riemann sphere induces a Möbius transformation of the complex plane. Rotations about the 3-axis are easy to compute. A rotation through an angle γ about the 3-axis is given by $R_3(\gamma)$ in (1.1); the corresponding Möbius transformation is simply $w = e^{i\gamma}z$. The corresponding matrix in $SL(2, C)$, however, is

$$\pm \begin{pmatrix} e^{i\gamma/2} & 0 \\ 0 & e^{-i\gamma/2} \end{pmatrix},$$

and not, for example $\begin{pmatrix} e^{i\gamma} & 0 \\ 0 & 1 \end{pmatrix}$ since this latter matrix is not in $SL(2, C)$.

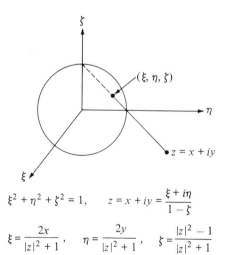

$$\xi^2 + \eta^2 + \zeta^2 = 1, \qquad z = x + iy = \frac{\xi + i\eta}{1 - \zeta}$$

$$\xi = \frac{2x}{|z|^2 + 1}, \qquad \eta = \frac{2y}{|z|^2 + 1}, \qquad \zeta = \frac{|z|^2 - 1}{|z|^2 + 1}$$

Figure 1.3. The stereographic projection.

The homomorphism

$$\begin{pmatrix} e^{i\gamma/2} & 0 \\ 0 & e^{-i\gamma/2} \end{pmatrix} \rightarrow \begin{pmatrix} \cos\gamma & -\sin\gamma & 0 \\ \sin\gamma & \cos\gamma & 0 \\ 0 & 0 & 1 \end{pmatrix}$$

shows immediately that $SO(3)$ is not simply connected. The path

$$\begin{pmatrix} e^{i\gamma/2} & 0 \\ 0 & e^{-i\gamma/2} \end{pmatrix}, \qquad 0 \le \gamma \le 2\pi$$

goes from $\mathbb{1}$ to $-\mathbb{1}$ in $SL(2, C)$, but it is mapped onto a closed path beginning and ending at $\mathbb{1}$ in $SO(3)$.

Next we rotate the Riemann sphere about the ξ and η axes. The $SO(3)$ matrices for these operations are given by $R_1(\alpha)$ and $R_2(\beta)$ in (1.1). We want to find the Möbius transformations corresponding to these rotations. One way is simply to insert the new values of ξ', η' and ζ' obtained by rotations into the formulae for the stereographic projections and work out the new z' in terms of α and β. A more "group theoretic" approach is this: A rotation by an angle β about the η-axis can be obtained by

(i) rotating the η-axis up to the ζ-axis using $R_1(\pi/2)$;
(ii) rotating about the ζ-axis by β; and
(iii) rotating the ζ-axis back to the η axis using $R_1(-\pi/2)$.

All we need to do this are the Möbius transformations for the rotations $R_1(\pm \pi/2)$ and those already found for $R_3(\beta)$. To find the Möbius transformation for $R_1(\pi/2)$, note that it must perform the following operations on $\mathbb{C}$: $0 \rightarrow i \rightarrow \infty \rightarrow -i \rightarrow 0$. We may easily write down a such a Möbius transformation, namely

$$w(z) = -i\frac{z+i}{z-i},$$

which corresponds to the $SL(2, C)$ matrix

$$\frac{1}{\sqrt{2}}\begin{pmatrix} 1 & i \\ i & 1 \end{pmatrix}$$

with inverse

$$\frac{1}{\sqrt{2}}\begin{pmatrix} 1 & -i \\ -i & 1 \end{pmatrix}.$$

The above sequence then gives

$$U_\eta(\beta) = \frac{1}{\sqrt{2}}\begin{pmatrix} 1 & -i \\ -i & 1 \end{pmatrix}\begin{pmatrix} e^{i\beta/2} & 0 \\ 0 & e^{-i\beta/2} \end{pmatrix}\frac{1}{\sqrt{2}}\begin{pmatrix} 1 & i \\ i & 1 \end{pmatrix}$$

$$= \begin{pmatrix} \cos\beta/2 & -\sin\beta/2 \\ \sin\beta/2 & \cos\beta/2 \end{pmatrix}.$$

So $\pm U_\eta(\beta) \rightarrow R_2(\beta)$ in $SO(3)$.

To find $U_\xi(\alpha)$ we perform in sequence $R_2(-\pi/2)$, $R_3(\alpha)$, and $R_2(\pi/2)$. The Möbius transformation giving $R_2(\pi/2)$ takes $\infty \to 1 \to 0 \to -1 \to \infty$. It is therefore given by

$$w(z) = \frac{z-1}{z+1},$$

with $SL(2, C)$ matrix

$$\frac{1}{\sqrt{2}}\begin{pmatrix} 1 & -1 \\ 1 & 1 \end{pmatrix}.$$

Performing the above sequence gives

$$U_\xi(\alpha) = \frac{1}{\sqrt{2}}\begin{pmatrix} 1 & -1 \\ 1 & 1 \end{pmatrix}\begin{pmatrix} e^{i\alpha/2} & 0 \\ 0 & e^{-i\alpha/2} \end{pmatrix}\frac{1}{\sqrt{2}}\begin{pmatrix} 1 & 1 \\ -1 & 1 \end{pmatrix}$$

and therefore

$$U_\xi(\alpha) = \begin{pmatrix} \cos\alpha/2 & i\sin\alpha/2 \\ i\sin\alpha/2 & \cos\alpha/2 \end{pmatrix}.$$

In summary, we have found the $SL(2, C)$ matrices that map onto rotations of the Riemann sphere ($SO(3)$ elements) about the coordinate axes

$$\pm U_\xi(\alpha) \to R_1(\alpha)$$
$$\pm U_\eta(\beta) \to R_2(\beta)$$
$$\pm U_\zeta(\gamma) \to R_3(\gamma).$$

These $SL(2, C)$ matrices are all unitary, so they belong to $SU(2)$, the 2×2 unitary matrices with determinant $+1$. The general $SU(2)$ matrix is a product of these and maps onto a product of $SO(3)$ matrices. Thus $SU(2)$ is a covering group of $SO(3)$. Now an arbitrary element of $SU(2)$ has the form

$$U = \begin{pmatrix} \alpha & \beta \\ -\bar{\beta} & \bar{\alpha} \end{pmatrix}, \qquad |\alpha|^2 + |\beta|^2 = 1.$$

Since α and β are complex, the manifold of $SU(2)$ is the sphere S^3 in $\mathbb{R}^4$. This is obviously a simply connected manifold; so $SU(2)$ is simply connected, and is therefore the universal covering group of $SO(3)$.

The unitary matrices found above can be written as exponentials of three famous matrices:

$$U_\xi(\alpha) = e^{i(\alpha/2)\sigma_1}$$
$$U_\eta(\beta) = e^{-i(\beta/2)\sigma_2} \tag{1.2}$$
$$U_\zeta(\gamma) = e^{i(\gamma/2)\sigma_3}$$

with

$$\sigma_1 = \begin{pmatrix} 0 & 1 \\ 1 & 0 \end{pmatrix}, \qquad \sigma_2 = \begin{pmatrix} 0 & -i \\ i & 0 \end{pmatrix}, \qquad \sigma_3 = \begin{pmatrix} 1 & 0 \\ 0 & -1 \end{pmatrix}. \qquad (1.3)$$

These are the *Pauli spin matrices* whose properties are developed in exercises below.

The annoying sign differences in the formulae (1.2) are the result of differing conventions used in the mathematics (stereographic projection) and physics (Pauli matrices) literatures. The covering homomorphism of $SO(3)$ by $SU(2)$ used by the physicists is obtained in an entirely different manner, as follows.

Let $\hat{n}$ be a unit vector and consider the matrix

$$U_{\hat{n}}(\theta) = \exp -\frac{i\theta}{2}\hat{n}\cdot\vec{\sigma} \qquad (1.4)$$

where $\hat{n}\cdot\vec{\sigma} = \sum_{j=1}^{3} n^j\sigma_j$. It is easily seen that $U_{\hat{n}}$ is a unitary matrix. Consider the following formulae:

$$e^{-i(\theta/2)\sigma_3}\sigma_1 e^{i(\theta/2)\sigma_3} = \sigma_1 \cos\theta + \sigma_2 \sin\theta$$

$$e^{-i(\theta/2)\sigma_3}\sigma_2 e^{i(\theta/2)\sigma_3} = -\sigma_1 \sin\theta + \sigma_2 \cos\theta \qquad (1.5)$$

$$e^{-i(\theta/2)\sigma_3}\sigma_3 e^{i(\theta/2)\sigma_3} = \sigma_3.$$

(The derivation of these formulae is left as an exercise.) Equations (1.5) show that the quantities σ_1, σ_2, and σ_3 transform as the components of a vector under a rotation about the 3-axis in the *clockwise* direction.

Rotations in $\mathbb{R}^3$ are obtained as follows. Given $X \in \mathbb{R}^3$ write

$$\mathfrak{X} = \sum_{j=1}^{3} x^j\sigma_j$$

and consider the operation

$$\mathfrak{X} \to \mathfrak{X}' = e^{-i\theta/2\hat{n}\cdot\vec{\sigma}}\mathfrak{X}e^{i\theta/2\hat{n}\cdot\vec{\sigma}}$$

$$= \sum_{j=1}^{3} x'^j\sigma_j.$$

Taking $\hat{n} = (0,0,1)$, for example, one finds that the quantities x^j transform as a rotation about the 3-axis in the counterclockwise sense. That is $x'^j = \sum_k (R_3(\theta))_{jk} x^k$ where $R_3(\theta)$ is given in (1.1). For a general $\hat{n}$ the $\sigma_1, \sigma_2, \sigma_3$ rotate about $\hat{n}$ in a clockwise direction through an angle θ while the components x^j rotate in the counterclockwise direction. The quantities x^j and σ_j are thus dual to each other.

For further properties of the Pauli spin matrices and the covering homomorphism, see the exercises.

We have found that two matrices in $SU(2)$, say U and $-U$, both map into the same rotation $R_U = R_{-U}$ in $SO(3)$. Moreover, a neighborhood of U and of $-U$ are each mapped continuously and one-to-one onto a neighborhood

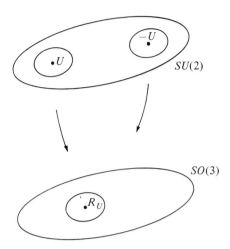

of R_U. Thus $SU(2)$ and $SO(3)$ are locally isomorphic. The representation of the element R of $SO(3)$ by the *pair* $\pm U$ in $SU(2)$ is called a "double-valued," "projective," "ray," or "spinor" representation of $SO(3)$. Such representations are significant in the quantum mechanics of particles with spin one-half, for their wavefunctions transform according to such "spinor representations." An electron, for example, is described by a two component wavefunction (spinor) $\binom{\psi_1}{\psi_2}$. Under a rotation R of three dimensional space—a geometrical transformation—the wavefunction is multiplied by a matrix U in $SU(2)$ which covers R. Thus

$$\begin{pmatrix}\psi_1\\\psi_2\end{pmatrix} \xrightarrow{R} U\begin{pmatrix}\psi_1\\\psi_2\end{pmatrix}.$$

This spinor representation is an ordinary single valued representation of $SU(2)$. In this sense, $SU(2)$ is the "quantum mechanical rotation group."

We have worked quite hard on $SU(2)$. Before leaving it we apply it to $SL(2, R)$. First consider the unit disk $|z| < 1$ in $\mathbb{C}$. A rotation about the ξ-axis by $\pi/2$ carries the disk into the upper half plane; an $SL(2, R)$ transformation then gives the most general isometric distortion of this half space; finally rotation by $-\pi/2$ about ξ brings the distorted half plane back to the disk. This sequence of transformations preserves the disk and is given by

$$\frac{1}{\sqrt{2}}\begin{pmatrix} 1 & -i \\ -i & 1 \end{pmatrix}\begin{pmatrix} a & b \\ c & d \end{pmatrix}\frac{1}{\sqrt{2}}\begin{pmatrix} 1 & i \\ i & 1 \end{pmatrix}$$

with a, b, c, d real. This matrix is of the form

$$\begin{pmatrix} \alpha & \beta \\ \bar{\beta} & \bar{\alpha} \end{pmatrix} \qquad \alpha, \beta \in C \tag{1.6}$$

with $|\alpha|^2 - |\beta|^2 = 1$. The set of all such matrices forms the group $SU(1,1)$ which, by construction, is isomorphic with $SL(2, R)$. The general definiton of the pseudo-unitary groups $SU(p,q)$ is given in §6. $SU(1,1)$ is the group of matrices U such that

$$U \begin{pmatrix} 1 & 0 \\ 0 & -1 \end{pmatrix} U^* = \begin{pmatrix} 1 & 0 \\ 0 & -1 \end{pmatrix}.$$

EXERCISES ON $SU(2)$

1. Check that $\sigma_i^* = \sigma_i$, $\operatorname{Tr} \sigma_i = 0$, $\sigma_i \sigma_j + \sigma_j \sigma_i = 2\delta_{ij}$, and $[\sigma_1, \sigma_2] = 2i\sigma_3$ and cyclically.

2. Use the results of exercise 1 to show that

$$\sigma_i \sigma_j = \delta_{ij} \mathbb{1} + i\varepsilon_{ijk} \sigma_k, \qquad \varepsilon_{ijk} = \begin{cases} 1 & ijk \text{ in cyclic order} \\ -1 & ijk \text{ in anti-cyclic order} \\ 0 & \text{if two indices are equal} \end{cases}$$

and more generally that

$$(\vec{a} \cdot \vec{\sigma})(\vec{b} \cdot \vec{\sigma}) = \vec{a} \cdot \vec{b} \mathbb{1} + i\vec{a} \times \vec{b} \cdot \vec{\sigma}$$

where $\vec{a} \cdot \vec{\sigma} = a^1 \sigma_1 + a^2 \sigma_2 + a^3 \sigma_3 = a^j \sigma_j$.
Show $e^{i(\theta/2)\hat{n} \cdot \vec{\sigma}} = \mathbb{1} \cos(\theta/2) + i\hat{n} \cdot \vec{\sigma} \sin(\theta/2)$.

3. Derive (5): Show that

$$e^{-i\gamma/2\sigma_3} \sigma_1 e^{+i\gamma/2\sigma_3} = \sigma_1 \cos \gamma + \sigma_2 \sin \gamma$$

$$e^{-i\gamma/2\sigma_3} \sigma_2 e^{+i\gamma/2\sigma_3} = -\sigma_1 \sin \gamma + \sigma_2 \cos \gamma$$

$$e^{-i\gamma/2\sigma_3} \sigma_3 e^{+i\gamma/2\sigma_3} = \sigma_3.$$

Thus $\vec{\sigma}$ is rotated through an angle $(-\gamma)$ by $e^{-i\gamma/2\sigma_3}$. This point is amplified in exercise 5. Hint: $e^{i\alpha\sigma_3} = \mathbb{1} \cos \alpha + i\sigma_3 \sin \alpha$. Use exercise 1.

4. $\mathbb{R}^3$ is mapped into the traceless Hermitian matrices by $\vec{X} \to \vec{X} \cdot \vec{\sigma}$. Show that $\det \vec{X} \cdot \vec{\sigma} = -(\vec{X} \cdot \vec{X})$, and that the transformation can be inverted by

$$x^i = \tfrac{1}{2} \operatorname{Tr}(\vec{X} \cdot \vec{\sigma} \sigma_i).$$

5. Compute $e^{-i(\theta/2)\hat{n} \cdot \vec{\sigma}} \vec{X} \cdot \vec{\sigma} e^{+i(\theta/2)\hat{n} \cdot \vec{\sigma}} = \vec{X}' \cdot \vec{\sigma}$ and show that

$$\vec{X}' = (\hat{n} \cdot \vec{X})\hat{n} + \cos \theta [(\hat{n} \times \vec{X}) \times \hat{n}] + \sin \theta [\hat{n} \times \vec{X}].$$

This is exactly the result of rotating $\vec{X}$ by an angle θ about the axis $\hat{n}$. The unitary matrix $U = e^{-i(\theta/2)\hat{n} \cdot \vec{\sigma}}$ rotates $\vec{\sigma}$ by $-\theta$ (exercise 3) and hence $\vec{X}$ by θ. We can also see this from the equation $\vec{X}' \cdot \vec{\sigma} = \Sigma(x)^j R_{ji} \sigma_i$: if $\vec{X}' = R\vec{X}$ then $\vec{\sigma} = R^t \vec{\sigma}$.

6. Show that the $SU(2)$ transformations $U_\xi(\alpha)$ and $U_\eta(\beta)$ become $SU(1,1)$ transformations when $\alpha \to i\alpha$ and $\beta \to i\beta$, or when $\sigma_1 \to -i\sigma_1$, or $\sigma_2 \to -i\sigma_2$.

7. Define $\vec{\tau} = (-i\sigma_1, -i\sigma_2, \sigma_3)$ and $\vec{X} \cdot \vec{\tau} = x^1 \tau_1 + x^2 \tau_2 + x^3 \tau_3$. Show that $\det \vec{X} \cdot \vec{\tau} = -(x^3)^2 + (x^1)^2 + (x^2)^2$ and find a homomorphism of $SU(1,1)$ onto $SO(2,1)$, the

group of real pseudo-orthogonal transformations of $\mathbb{R}^3$ that preserve the metric $(x^1)^2 + (x^2)^2 - (x^3)^2$.

8. Let K be the subgroup of $SU(1,1)$ consisting of matrices of the form $\begin{pmatrix} e^{i\theta} & 0 \\ 0 & e^{-i\theta} \end{pmatrix}$.

Show that the cosets $SU(1,1)/K$ are in one-to-one correspondence with the unit disk $|z| < 1$. Show $SU(1,1)$ is not simply connected. Show that the metric

$$\frac{dz \, d\bar{z}}{(1 - |z|^2)^2}$$

is invariant under the action of $SU(1,1)$ on the unit disk.

5. The Lorentz Group

Space-time in special relativity, or *Minkowski space*, is $\mathbb{R}^4$ with the metric

$$(X, X) = (x^0)^2 - \vec{X} \cdot \vec{X}$$

where $x^0 = ct$ with c the speed of light and $\vec{X}$ is an ordinary vector in $\mathbb{R}^3$. $X = (x^0, \vec{X})$ is called a four-vector. The Lorentz transformations are those linear transformations of Minkowski space that leave (X, X) invariant: $X \to X' = \Lambda X$ with

$$(\Lambda X, \Lambda X) = (X, X).$$

We can write this as $(\Lambda X)^t g(\Lambda X) = X^t g X$ with

$$g = \begin{pmatrix} 1 & & & \\ & -1 & & \\ & & -1 & \\ & & & -1 \end{pmatrix}.$$

So Λ is a Lorentz transformation if

$$\Lambda^t g \Lambda = g. \tag{1.7}$$

The set of such Λ is $O(3, 1)$, a real Lie group.

The Lorentz group is the union of four disconnected pieces. From (1.7) we find that $(\det \Lambda)^2 = 1$ and also

$$\Lambda_{00}^2 - (\Lambda_{10}^2 + \Lambda_{20}^2 + \Lambda_{30}^2) = 1. \tag{1.8}$$

It follows that $\Lambda_{00}^2 \geq 1$. The four components are thus labeled by $\det \Lambda = \pm 1$ and $\operatorname{sgn} \Lambda_{00} = \pm 1$. We denote by $SO(3, 1)^{\uparrow}$ the subgroup of $O(3, 1)$ for which $\det \Lambda = 1$ and $\operatorname{sgn} \Lambda_{00} = 1$. This subgroup is called the *proper orthochronous* Lorentz group. Its elements preserve parity (spatial orientation) and the direction of time (this is indicated by the arrow.). Further examination shows that $SO(3, 1)^{\uparrow}$ is actually a normal subgroup of $O(3, 1)$. This is most easily seen by observing that it is the kernel of the homomorphism

$$\Lambda \to (\det \Lambda, \operatorname{sgn} \Lambda_{00})$$

from $O(3, 1)$ into the multiplicative group consisting of four elements $\{(a, b) | a = \pm 1, b = \pm 1\}$. (Multiplication in this group is defined by $(a, b) \circ (a', b') = (aa', bb')$.)

The quotient group $O(3, 1)/SO(3, 1)^\uparrow$ has four elements. As coset representatives we may take the matrices

$$\mathbb{1}, \quad \Lambda_P = \begin{pmatrix} 1 & & & \\ & -1 & & \\ & & -1 & \\ & & & -1 \end{pmatrix}, \quad \Lambda_T = \begin{pmatrix} -1 & & & \\ & 1 & & \\ & & 1 & \\ & & & 1 \end{pmatrix},$$

$$\Lambda_{PT} = \begin{pmatrix} -1 & & & \\ & -1 & & \\ & & -1 & \\ & & & -1 \end{pmatrix}.$$

Thus $O(3, 1)/SO(3, 1)^\uparrow$ is isomorphic to the group $\{\mathbb{1}, \Lambda_P, \Lambda_T, \Lambda_{PT}\}$. The matrix Λ_P performs a spatial inversion (P stands for parity); while Λ_T reverses time and Λ_{PT} is a total inversion. The corresponding coset decomposition represents the decomposition of $O(3, 1)$ into the union of four connected components:

$$O(3, 1) = \mathbb{1} SO(3, 1)^\uparrow \cup \Lambda_P SO(3, 1)^\uparrow \cup \Lambda_T SO(3, 1)^\uparrow \cup \Lambda_{PT}(3, 1)^\uparrow.$$

We now focus on $SO(3, 1)^\uparrow$ and show that its universal covering group is $SL(2, C)$. We begin by representing points in Minkowski space by 2×2 matrices. Let σ_0 be the 2×2 identity matrix and consider

$$(x^0, \vec{X}) \to \mathfrak{X} = x^\mu \sigma_\mu = \begin{pmatrix} x^0 + x^3 & x^1 - ix^2 \\ x^1 + ix^2 & x^0 - x^3 \end{pmatrix}.$$

(Again, σ_1, σ_2, and σ_3 are the Pauli spin matrices.) This mapping is invertible, and in fact

$$x^\mu = \tfrac{1}{2} \operatorname{Tr} \mathfrak{X} \sigma_\mu.$$

Moreover, the Minkowski inner product is given by

$$(X, X) = \det \mathfrak{X}.$$

Now consider the transformation $\mathfrak{X} \to A\mathfrak{X}A^*$ with A in $SL(2, C)$. Since $\det A = 1$ this transformation preserves (X, X). Writing

$$\mathfrak{X}' = A\mathfrak{X}A^* = x'^\mu \sigma_\mu$$

we see that there is a Lorentz transformation Λ_A such that $X' = \Lambda_A X$. It is clear that in this identification we must have $\Lambda_A = \Lambda_{-A}$ so there is a two-to-one homomorphism of $SL(2, C)$ into $SO(3, 1)^\uparrow$. We leave it as a series of exercises to show that this homomorphism is onto, that $\pm A \to \Lambda_A$ is continuous and locally one-to-one, and that $SL(2, C)$ is simply connected. Thus $SL(2, C)$ is the universal covering group of $SO(3, 1)^\uparrow$.

The physical significance of $SL(2, C)$ is that particles transform according to its representations: when we perform a Lorentz transformation on space-time, the wavefunctions (spinors, vectors, tensors, etc.) that describe the particles are multiplied by one of the matrices in a representation of $SL(2, C)$ that covers the Lorentz transformation.

EXERCISES ON THE LORENTZ GROUP

1. Show that $\Lambda \to (\det \Lambda, \Lambda_{00})$ is a homomorphism. Prove that the connected component of the identity in any Lie group is a normal subgroup. Show that the mapping $A \to \Lambda_A$ is a homomorphism.

2. Show that for any complex χ,

$$A = e^{(\chi/2)\hat{n} \cdot \vec{\sigma}}$$

is in $SL(2, C)$. Show that Λ_A is a rotation about the $\hat{n}$ axis for $\chi = i\theta$. Compute Λ_A for real χ and $\hat{n} = (0, 0, 1)$; this is a pure Lorentz "boost" along the 3-axis.

3. Try to find A for Λ_P. It won't work. Repeat this attempt for Λ_{PT} and Λ_T. This shows that $SL(2, C)$ only covers $SO(3, 1)^\uparrow$ and not any other piece of $O(3, 1)$.

4. Show that $SL(2, C)$ is simply connected. Use the polar decomposition of a matrix $A \in SL(2, C)$: $A = HU$, with $U^* = U^{-1}$, $H^* = H$, $\det H = \det U = 1$, and $\operatorname{Tr} H > 0$. (You may wish to prove that this decomposition is correct. To see this, note that AA^* is a positive Hermitian matrix so can be diagonalized by a unitary matrix $P: PAA^*P^* = D^2$. Let D be the positive square root of D^2, so $\operatorname{Tr} D > 0$. Then $H = P^*DP$ and $U = H^{-1}M$.) A closed path in $SL(2, C)$ is $A(t) = H(t)U(t), 0 \le t \le 1$ with $A(0) = A(1)$. Show $U(0) = U(1)$ and $H(0) = H(1)$. We have already seen that any closed path in $SU(2)$ can be shrunk to a point, so investigate the manifold of the Hermitian matrices with $\operatorname{Tr} H > 0$. Show that it is a simply connected smooth, non-compact manifold.

5. How many disconnected components does $O(3, 2)$ have? This is one of the de Sitter groups. $\Lambda \in O(3, 2) \Leftrightarrow \Lambda^t g \Lambda = g$ with $g = \begin{pmatrix} 1 & & & & \\ & 1 & & & \\ & & -1 & & \\ & & & -1 & \\ & & & & -1 \end{pmatrix}$.

6. Show that $GL(2, C)$ is not simply connected. What is its covering group?

6. The Classical Groups

The last examples we introduce here are "the classical groups." These are subgroups of $GL(n, C)$—"her all embracing majesty" (Weyl [2])—the group of non-singular linear transformations of an n-dimensional complex vector space. We will treat this as the group of $n \times n$ complex matrices with non-zero determinant. $GL(n, C)$ has n^2 complex parameters (therefore $2n^2$ real ones). The homomorphism $\rho: A \to \det A$ has for its kernel the normal subgroup $SL(n, C)$, the group of $n \times n$ complex matrices with determinant 1. $SL(n, C)$ has $n^2 - 1$ parameters; it is called A_{n-1} in Cartan's notation (discussed in Chapter 2).

The other classical groups leave certain bilinear forms on the vector space invariant. First consider the *symmetric* bilinear form

$$(x, y) = \sum_{i=1}^{n} x^i y^i \equiv x^t y.$$

The matrices which leave this form invariant make up the orthogonal group $O(n)$. They satisfy

$$A^t A = 1.$$

These are $n(n + 1)/2$ equations constraining the n^2 entries of A, so $O(n)$ is an $n(n - 1)/2$ parameter group. The constraints do not bound the entries of A in $O(n, C)$, so the group is not compact. If $n = 2k + 1$, then it is called B_k in Cartan's notation; if $n = 2k$, it is D_k.

Next consider the skew-symmetric bilinear form in a (necessarily even) $2n$ dimensional vector space

$$(x, y) = (x^1 y^{n+1} - x^{n+1} y^1) + (x^2 y^{n+2} - x^{n+2} y^2) + \cdots + (x^n y^{2n} - x^{2n} y^n)$$

$$= x^t J y$$

with

$$J = \begin{pmatrix} 0 & 1_n \\ -1_n & 0 \end{pmatrix}, \qquad 1_n = n \times n \text{ unit matrix.}$$

Matrices which leave this form invariant satisfy

$$A^t J A = J$$

and constitute the non-compact symplectic group $Sp(2n)$, or C_n in Cartan's notation.

When dealing with a complex vector space we can consider the Hermitian symmetric forms

$$(x, y) = \sum_{i=1}^{n} \overline{x^i} y^i \equiv x^* y$$

or more generally

$$(x, y) = x^* g y$$

with

$$g = \begin{pmatrix} 1_p & 0 \\ 0 & -1_q \end{pmatrix}$$

where $p + q = n$. These forms remain invariant under transformations that satisfy

$$A*gA = g.$$

Those A for which $\det A = 1$ form the pseudo-unitary groups $SU(p, q)$. We will see in the next chapter that they have the same complex Lie algebra as $SL(p + q)$, so these are also A_{n-1} in Cartan's notation. If $q = 0$ the condition $A*A = 1$ bounds all the entries by 1, so $SU(n)$ is compact.

Finally, the reader should verify that the skew-Hermitian form $(x, y) = x*Jy$ in an even dimensional complex space is equivalent to one of the above bilinear forms, so it yields no new classical group.

Each of these classical groups becomes a real group if the matrix elements are restricted to be real instead of complex—$SL(n)$ becomes $(SL(n, R)$ and $SU(n)$ becomes $SO(n, R)$, for example—but there are other ways to associate a real group with a complex group. We will discuss the real groups more fully later.

The Classical Groups and Their Invariant Forms

Group	Form	Comments	Cartan Family
$SL(n)$	—	$\det A = 1$	A_{n-1}
$SO(2n + 1)$	$x^t y$	Symmetric form	B_n
$SO(2n)$	$x^t y$		D_n
$Sp(2n)$	$x^t J y$	$J = \begin{pmatrix} 0 & 1_n \\ -1_n & 0 \end{pmatrix}$ Skew-symmetric form	C_n
$SU(p, q)$	$x*gy$	$g = \begin{pmatrix} 1_p & 0 \\ 0 & -1_q \end{pmatrix}$ Hermitian form, indefinite if $q \neq 0$	A_{n-1}

EXERCISES ON THE CLASSICAL GROUPS

It is often useful to write a typical matrix in $SU(p, q)$ or $Sp(2n)$ as $\begin{pmatrix} A & B \\ C & D \end{pmatrix}$ where the submatrices have appropriate dimensionality, e.g. B has p rows and q columns in $SU(p, q)$.

1. Show that the complex groups below depend on the number of *real* parameters shown:

$SL(n)$	$2n^2 - 1$
$SO(n)$	$n(n - 1)$
$SU(p, q)$	$(p + q)^2 - 1$
$Sp(2n)$	$n(n + 1)$

2. What conditions must the $n \times n$ matrices A, B, C and D obey so that $M = \begin{pmatrix} A & B \\ C & D \end{pmatrix}$ leaves *both* the forms $x^t y$ and $x*Jy$ invariant? This group is called $SO*(2n)$.

CHAPTER 2

Lie Algebras

7. Real and Complex Lie Algebras

We have said that a Lie group is an analytic manifold; and so it makes sense to talk about the tangent space to that manifold, and in particular the tangent space at the identity of the group. That tangent space is called the Lie algebra. In the case of a linear group it can be computed explicitly by differentiating curves through the identity.

For example, the rotations (1.1) are all curves through the identity in $SO(3)$. Their derivatives at the origin, $L_j = \dot{R}_j(0)$, are given by

$$L_1 = \begin{pmatrix} 0 & 0 & 0 \\ 0 & 0 & -1 \\ 0 & 1 & 0 \end{pmatrix} \quad L_2 = \begin{pmatrix} 0 & 0 & 1 \\ 0 & 0 & 0 \\ -1 & 0 & 0 \end{pmatrix} \quad L_3 = \begin{pmatrix} 0 & -1 & 0 \\ 1 & 0 & 0 \\ 0 & 0 & 0 \end{pmatrix}. \quad (2.1)$$

These matrices form the basis for the Lie algebra $so(3)$. They in turn generate the one-parameter subgroups $R_j(\alpha)$ by the exponential formula $R_j(\alpha) = e^{\alpha L_j}$. We say that the matrices L_j are the infinitesimal generators of the Lie group $SO(3)$.

As another example let us compute the Lie algebra $su(2)$. The group $SU(2)$ consists of 2×2 matrices U such that $UU^* = I$ and $\det U = 1$. We let $U(t)$ be a curve passing through the identity at $t = 0$, and differentiate these identities at $t = 0$. We obtain

$$\frac{dUU^*}{dt}\bigg|_{t=0} = \dot{U} + \dot{U}^* = 0,$$

$$\frac{d \det U}{dt}\bigg|_{t=0} = \operatorname{Tr} \dot{U} = 0.$$

The algebra $su(2)$ therefore consists of all 2×2 skew Hermitian matrices of trace zero.

As a third example, consider the Lorentz group $SO(3, 1)$. The matrices Λ satisfy $\Lambda^t g \Lambda = g$, where g is the Minkowski metric. (See 1.7). If $\Lambda(t)$ is a curve passing through the identity at $t = 0$ we have

$$\frac{d\Lambda^t g\Lambda}{dt}\bigg|_{t=0} = L^t g + gL = 0, \qquad L = \dot{\Lambda}(0).$$

The Lie algebra $so(3, 1)$ is given by $\{L | L^t g + gL = 0\}$.

A Lie algebra is a linear vector space, but because of the group structure on the manifold it inherits a rich algebraic structure. In this chapter we shall focus on Lie algebras as entities in themselves and discuss some of the most important aspects of their structure. A deeper investigation of their structure will be deferred until Chapters 9 and 10.

A Lie algebra $\mathfrak{g}$ is a vector space over a field F on which a product $[\ ,\ ]$, called the Lie bracket, is defined, with the properties

(1) $X, Y \in \mathfrak{g}$ imply $[X, Y] \in \mathfrak{g}$
(2) $[X, \alpha Y + \beta Z] = \alpha[X, Y] + \beta[X, Z]$
   for $\alpha, \beta \in F$ and $X, Y, Z \in \mathfrak{g}$
(3) $[X, Y] = -[Y, X]$
(4) $[X, [Y, Z]] + [Y, [Z, X]] + [Z, [X, Y]] = 0.$

Property 3 is called skew symmetry and 4 is known as the Jacobi identity. If F is the field of real numbers $\mathbb{R}$ we say $\mathfrak{g}$ is a real Lie algebra; and we say $\mathfrak{g}$ is complex if $F = \mathbb{C}$. These are the only two cases we shall consider in this book.

A *matrix* or *linear* Lie algebra is an algebra of matrices with the commutator $XY\text{-}YX$ taken as the Lie bracket $[X, Y]$. Note that the commutator satisfies properties 2 through 4. For example, the set of all $n \times n$ matrices with entries in F is a Lie algebra, known as $gl(n, F)$. The set of skew-Hermitian matrices $(A^* = -A)$, forms a Lie algebra over $\mathbb{R}$ known as $su(n)$. (Note that the algebra itself is real, because the scalars α, β in property 2 are real, even though the matrices themselves may have complex entries.) The set of all matrices of trace zero with entries in F forms a Lie algebra, known as $sl(n, F)$.

If $\{E_i\}$ is a basis for $\mathfrak{g}$ then we must have

$$[E_i, E_j] = \sum_k C_{ij}^k E_k$$

for some set of constants C_{ij}^k, called the *structure constants* of the algebra. Accordingly, a Lie algebra may be specified by giving a set of constants C_{ij}^k such that

$$C_{ij}^k = -C_{ji}^k$$

$$C_{ij}^m C_{mk}^r + C_{jk}^m C_{mi}^r + C_{ki}^m C_{mj}^r = 0, \qquad r = 1, \ldots, n.$$

Here we have used the *summation convention*: We sum over a repeated index (in this case, m).

A simple example of a Lie algebra is furnished by $\mathbb{R}^3$ with the usual cross product as the Lie bracket: $[X, Y] = X \times Y$. If (E_1, E_2, E_3) is a right handed orthonormal basis for $\mathbb{R}^3$ then

$$E_j \times E_k = \varepsilon_{jkl} E_l$$

where ε_{jkl} is the completely anti-symmetric tensor. The components ε_{jkl} are thus the structure constants of this Lie algebra.

A second example of a Lie algebra is the real algebra of matrices (2.1) with the commutator as Lie bracket. The commutators of the $\{L_i\}$ are

$$[L_j, L_k] = \varepsilon_{jkl} L_l.$$

This algebra is known as $so(3)$; it is the Lie algebra of the rotation group.

These two algebras are quite clearly the same. We say two Lie algebras $\mathfrak{g}_1$ and $\mathfrak{g}_2$ are *isomorphic* if there is a linear invertible mapping ρ from $\mathfrak{g}_1$ to $\mathfrak{g}_2$ such that

$$\rho(\alpha X + \beta Y) = \alpha \rho(X) + \beta \rho(Y)$$

$$[\rho(X), \rho(Y)]_2 = \rho([X, Y]_1),$$

where $[\ ,\ ]_1$ and $[\ ,\ ]_2$ denote the brackets in $\mathfrak{g}_1$ and $\mathfrak{g}_2$ respectively. We say that $\mathfrak{g}_1$ and $\mathfrak{g}_2$ are isomorphic as real Lie algebras if and only if the mapping ρ is real, that is, only if the matrix of the linear transformation ρ is real.

A third example, isomorphic to the two above, is the real Lie algebra $su(2)$ spanned by

$$E_1 = \frac{1}{2}\begin{pmatrix} 0 & -i \\ -i & 0 \end{pmatrix}, \qquad E_2 = \frac{1}{2}\begin{pmatrix} 0 & -1 \\ 1 & 0 \end{pmatrix}, \qquad E_3 = \frac{1}{2}\begin{pmatrix} -i & 0 \\ 0 & i \end{pmatrix}. \quad (2.2)$$

These matrices are precisely $E_j = -\frac{1}{2} i \sigma_j$ where σ_j are the Pauli spin matrices of Chapter 1. In the physics literature the matrices $\frac{1}{2} \sigma_j$ are used, since these are Hermitian; but any element X in the real Lie algebra $su(2)$ can be written with real coefficients x^j:

$$X = \sum_j x^j (-\tfrac{i}{2}\sigma_j) = \sum_j (-ix^j)(\tfrac{1}{2}\sigma_j).$$

Thus, when the σ_j are used as a basis, the coefficients are purely imaginary.

The set of 2×2 complex matrices with trace zero forms the complex Lie algebra $sl(2, C)$. Since the matrices (2.2) have trace zero they lie in $sl(2, C)$, and in fact form a basis for $sl(2, C)$ over the complex numbers. Thus, $sl(2, C)$ may be regarded as the *complexification* of the real Lie algebra $su(2)$ obtained by taking linear combinations of the matrices in $su(2)$ with complex coefficients. Another basis for $sl(2, C)$ is given by

$$a_+ = \begin{pmatrix} 0 & 1 \\ 0 & 0 \end{pmatrix} \qquad a_- = \begin{pmatrix} 0 & 0 \\ 1 & 0 \end{pmatrix} \qquad a_0 = \frac{1}{2}\begin{pmatrix} 1 & 0 \\ 0 & -1 \end{pmatrix}.$$

The commutation relations of these matrices are

$$[a_0, a_\pm] = \pm a_\pm, \qquad [a_+, a_-] = 2a_0.$$

The real Lie algebra generated by a_0, a_+, a_- is $sl(2, R)$. The real Lie algebras $su(2)$ and $sl(2, R)$ have the same complexification, $sl(2, C)$; but we shall see in Chapter 11 that they are not isomorphic as real algebras. That is, there is no isomorphism from one to the other involving only real coefficients. These two algebras are thus distinct real forms of the complex Lie algebra $sl(2, C)$.

There is, however, a complex isomorphism from $su(2)$ to $sl(2, R)$, namely

$$a_\pm = \frac{\sigma_1 \pm i\sigma_2}{2} \qquad a_0 = \frac{1}{2}\sigma_3.$$

On the other hand, the real Lie algebra with basis $\{b_+, b_-, b_0\}$ and commutators

$$[b_0, b_\pm] = \pm b_\pm, \qquad [b_+, b_-] = -2b_0 \qquad (2.3)$$

is isomorphic to $sl(2, R)$. We leave it as an exercise to find the real isomorphism.

A real Lie algebra $\mathfrak{h}$ is said to be a real form of the complex algebra $\mathfrak{g}$ if $\mathfrak{g}$ is the complexification of $\mathfrak{h}$. Two real forms are distinct if there is no real isomorphism between them. In the case above, $sl(2, C)$ has two distinct real forms, $su(2)$ and $sl(2, R)$; all other real forms of $sl(2, C)$ are isomorphic to one of these two. The subject of real forms will be discussed in greater detail in Chapter 11.

8. Representations of Lie Algebras

Lie algebras are usually realized in applications as operators on vector spaces with the commutator as the Lie bracket. For example, let

$$R_k = x^j \frac{\partial}{\partial x^i} - x^i \frac{\partial}{\partial x^j} \qquad i, j, k = 1, 2, 3 \qquad (2.4)$$

where i, j, k are in cyclic order. Then

$$[R_j, R_k] = \varepsilon_{jkl} R_l.$$

These differential operators act, for example, on the infinite dimensional vector space $C^\infty(R^3)$, and so provide an infinite dimensional *representation* of $so(3)$. A *representation* of a Lie algebra $\mathfrak{g}$ on a vector space V is a mapping ρ from $\mathfrak{g}$ to the linear transformations of V such that

$$\rho(\alpha X + \beta Y) = \alpha\rho(X) + \beta\rho(Y)$$

$$\rho([X, Y]_\mathfrak{g}) = [\rho(X), \rho(Y)] = \rho(X)\rho(Y) - \rho(Y)\rho(X),$$

$[\ ]_\mathfrak{g}$ being the Lie bracket of the algebra $\mathfrak{g}$. The dimension of the representation is equal to the dimension of V.

In Chapter 6 we shall discuss operators that generate the symmetries of differential equations. The symmetries of the heat equation, $u_t = u_{xx}$, for example, are generated by

$$X_1 = \frac{\partial}{\partial x} \qquad X_2 = \frac{\partial}{\partial t} \qquad X_3 = u\frac{\partial}{\partial u}$$

$$X_4 = x\frac{\partial}{\partial x} + 2t\frac{\partial}{\partial t} \qquad X_5 = 2t\frac{\partial}{\partial x} - xu\frac{\partial}{\partial u}$$

$$X_6 = 4tx\frac{\partial}{\partial x} + 4t^2\frac{\partial}{\partial t} - (x^2 + 2t)u\frac{\partial}{\partial u}.$$

These operators form a six dimensional Lie algebra, as one may see by computing their commutators:

	X_1	X_2	X_3	X_4	X_5	X_6
X_1	0	0	0	X_1	$-X_3$	$2X_5$
X_2	0	0	0	$2X_2$	$2X_1$	$2X_4 - 2X_3$
X_3	0	0	0	0	0	0
X_4	$-X_1$	$-2X_2$	0	0	X_5	$2X_6$
X_5	X_3	$-2X_1$	0	$-X_5$	0	0
X_6	$-2X_5$	$-4X_4 + 2X_3$	0	$-2X_6$	0	0

Another six dimensional Lie algebra is obtained by adding the differential operators $P_j = \partial/\partial x^j$ to the operators R_1, R_2, R_3 in (2.4). The commutators of the resulting algebra are

$$[R_j, R_k] = \varepsilon_{jkl} R_l$$

$$[R_j, P_k] = \varepsilon_{jkl} P_l$$

$$[P_j, P_k] = 0.$$

This algebra, called $e(3)$ generates the group of Euclidean motions in the plane.

Another example is furnished by the operators $Q = x$ and $P = \partial/\partial x$ acting on functions defined on the real line. The operator Q is multiplication by x; thus $(Qf)(x) = xf(x)$. The commutation relations are

$$[P, Q] = 1.$$

Introducing

$$a = \frac{P + Q}{\sqrt{2}}, \qquad a^* = \frac{Q - P}{\sqrt{2}}$$

we have $[a, a^*] = 1$. The Lie algebra $\{P, Q, 1\}$ is called the *Heisenberg algebra* and plays a fundamental role in quantum mechanics (see Chapter 4).

9. Some Important Concepts and Examples in Lie Algebras

(A) We say that $\mathfrak{s}$ is a *subalgebra* of $\mathfrak{g}$ if $\mathfrak{s}$ is closed under commutation, that is, if $[s_1, s_2] \in \mathfrak{s}$ whenever s_1 and s_2 belong to $\mathfrak{s}$. We often abbreviate this by writing $[\mathfrak{s}, \mathfrak{s}] \subset \mathfrak{s}$. In $e(3)$, for example, the algebras $\{R_1, R_2, R_3\}$ and $\{P_1, P_2, P_3\}$ are subalgebras called $so(3)$ and $t(3)$. The elements R_i and P_i are said to generate rotations and translations respectively.

(B) An algebra $\mathfrak{g}$ is the *direct sum* of two algebras $\mathfrak{a}$ and $\mathfrak{b}$ if $\mathfrak{g} = \mathfrak{a} + \mathfrak{b}$ as a vector space, and if $[\mathfrak{a}, \mathfrak{b}] = 0$. The algebras $\mathfrak{a}$ and $\mathfrak{b}$ are clearly subalgebras. For example, $t(3)$ is the direct sum of the three subalgebras $\{P_1\}$, $\{P_2\}$ and $\{P_3\}$. We write $\mathfrak{g} = \mathfrak{a} \oplus \mathfrak{b}$ to indicate a direct sum.

(C) An algebra $\mathfrak{g}$ is the *semi-direct sum* of two subalgebras $\mathfrak{a}$ and $\mathfrak{b}$ if $\mathfrak{g} = \mathfrak{a} + \mathfrak{b}$ as a vector space, but $[\mathfrak{a}, \mathfrak{b}] \subset \mathfrak{a}$. In this case we write $\mathfrak{g} = \mathfrak{a} \oplus_s \mathfrak{b}$. The algebra $e(3)$ is the semi-direct sum of $t(3)$ and $so(3)$. The algebra of the heat equation is the semi-direct sum $\{X_1, X_3, X_5\} \oplus_s \{X_2, X_4 - \frac{1}{2}X_3, X_6\}$.

(D) A subalgebra $\mathfrak{s}$ is an *ideal* of $\mathfrak{g}$ if $[\mathfrak{s}, \mathfrak{g}] \subset \mathfrak{s}$, that is, if $[A, B] \in \mathfrak{s}$ whenever $A \in \mathfrak{s}$ and $B \in \mathfrak{g}$. Thus $t(3)$ is an ideal of $e(3)$, but $so(3)$ is not. More generally, if $\mathfrak{g} = \mathfrak{a} \oplus_s \mathfrak{b}$ then $\mathfrak{a}$ is an ideal of $\mathfrak{g}$. In a direct sum, $\mathfrak{g} = \mathfrak{a}_1 \oplus \mathfrak{a}_2 \oplus \cdots \oplus \mathfrak{a}_n$, each of the summands $\mathfrak{a}_i$ is an ideal of $\mathfrak{g}$.

(E) The *center of* $\mathfrak{g}$ is the largest ideal $\mathfrak{c}$ such that $[\mathfrak{g}, \mathfrak{c}] = 0$. It is unique. The center of $e(3)$ is zero, while $\{P_3\}$ is the center of the subalgebra $\mathfrak{g}$ of $e(3)$ consisting of R_3, P_1, P_2, P_3.

(F) Just as there is a notion of quotient group in the theory of groups there is a notion of *quotient algebra* in the theory of Lie algebras. Recall that if H is a subgroup of G we define an equivalence relation on G by $a \equiv b(\mathrm{mod}\, H)$ if $a^{-1}b \in H$. The equivalence classes under this relation are called the left cosets of H and are denoted by aH. Similarly, we may define a second equivalence relation by $a \equiv b(\mathrm{mod}\, H)$ if $ab^{-1} \in H$; the equivalence classes in this case are the right cosets of H, denoted by Ha. We say H is normal if $aH = Ha$ for all $a \in G$. In that case the cosets of H in G form a group, with the group operation defined by $(aH)(bH) = abH$. The fact that H is normal is used to prove that the operation is well defined. This group is called the quotient group and is denoted by G/H.

Now suppose $\mathfrak{s}$ is a subalgebra of the Lie algebra $\mathfrak{g}$. For any $X \in \mathfrak{g}$ define $X + \mathfrak{s}$ to be the equivalence class of X under the equivalence relation $X \equiv Y(\mathrm{mod}\, \mathfrak{s})$ if $X - Y \in \mathfrak{s}$. In general these equivalence classes do not form a Lie algebra, but they do if $\mathfrak{s}$ is an ideal. In that case we define a Lie bracket on the classes by

$$[X + \mathfrak{s}, Y + \mathfrak{s}] = [X, Y] + \mathfrak{s}.$$

This bracket is well defined because $\mathfrak{s}$ is an ideal. The set of equivalence classes thus forms a new Lie algebra called the *quotient algebra*. The quotient algebra is denoted by $\mathfrak{g}/\mathfrak{s}$. For example, $e(3)/t(3)$ is a Lie algebra which is isomorphic to $so(3)$. The analogous group theoretic fact is that the quotient of the Euclidean group by the translations is the rotation group. In fact, ideals of Lie algebras always correspond to normal subgroups of the corresponding Lie group.

(G) The set of commutators $[\mathfrak{g}, \mathfrak{g}]$ is an ideal of $\mathfrak{g}$ (prove it), called $\mathfrak{g}^{(1)}$. Similarly $\mathfrak{g}^{(2)} = [\mathfrak{g}^{(1)}, \mathfrak{g}^{(1)}]$ is an ideal of $\mathfrak{g}^{(1)}$. We define $\mathfrak{g}^{(n+1)} = [\mathfrak{g}^{(n)}, \mathfrak{g}^{(n)}]$. If this sequence terminates in zero, we say $\mathfrak{g}$ is *solvable*. For example $e(3)^{(1)} = e(3)$, so $e(3)^{(n)} = e(3)$ and $e(3)$ is not solvable. On the other hand the subalgebra $\mathfrak{q} = \{R_3, P_1, P_2, P_3\}$ introduced above is solvable because $\mathfrak{q}^{(2)}$ is zero. The set of all $n \times n$ upper triangular matrices

$$\begin{pmatrix} a_{11} & a_{12} & \cdots & a_{1n} \\ 0 & a_{22} & \cdots & a_{2n} \\ \vdots & & \ddots & \\ 0 & \cdots & & a_{nn} \end{pmatrix} \tag{2.5}$$

is another solvable Lie algebra. Conversely, Lie proved that every complex solvable matrix algebra is isomorphic to a subalgebra of triangular matrices.

(H) Now consider a different sequence of ideals given by $\mathfrak{g}_{(1)} = [\mathfrak{g}, \mathfrak{g}]$, $\mathfrak{g}_{(2)} = [\mathfrak{g}, \mathfrak{g}_{(1)}], \ldots, \mathfrak{g}_{(n+1)} = [\mathfrak{g}, \mathfrak{g}_{(n)}]$. This is a nested sequence with $\mathfrak{g}_{(n+1)} \subseteq \mathfrak{g}_{(n)} \subseteq \cdots \subseteq \mathfrak{g}_{(1)} = \mathfrak{g}^{(1)} \subseteq \mathfrak{g}$. We say $\mathfrak{g}$ is *nilpotent* if this sequence terminates in zero. An important example of a nilpotent algebra is the Heisenberg algebra $\mathfrak{h} = \{P, Q, \mathbb{1}\}$.

It is easily seen that $\mathfrak{g}_{(n)} \supseteq \mathfrak{g}^{(n)}$, so that nilpotency implies solvability. On the other hand the solvable algebra (2.5) is not nilpotent. Another example of a solvable algebra which is not nilpotent is the algebra $\mathfrak{q}$ in (G) above. We find $\mathfrak{q}_{(1)} = \{P_1, P_2\} = \mathfrak{q}_{(n)}$ for all $n \geq 1$. The reader may show that the matrices

$$\begin{pmatrix} \lambda & a_{12} & \cdots & a_{1n} \\ 0 & \lambda & & \vdots \\ \vdots & & \ddots & \\ 0 & & & \lambda \end{pmatrix} \tag{2.6}$$

form a nilpotent algebra.

(I) The *radical R* of a Lie algebra is the maximal solvable ideal. The existence of the radical will be proved in Chapter 9. It is unique and contains all other solvable ideals. For example, $\mathfrak{t}(3)$ is the radical of $e(3)$.

A Lie algebra $\mathfrak{g}$ is *simple* if it contains no ideals other than $\mathfrak{g}$ and $\{0\}$; it is *semi-simple* if it contains no abelian ideals (other than $\{0\}$). Levi's decomposition (Chapter 9) states that every Lie algebra is the semi-direct sum of its radical and a semi-simple Lie algebra. For example, $e(3) = \mathfrak{t}(3) \oplus_s so(3)$ is the Levi decomposition of $e(3)$.

The semi-simple Lie algebras constitute an important class of Lie algebras and play a fundamental role in geometry and physics. A complete list and classification of the complex semi-simple Lie algebras was given by Cartan. We shall give that list at the end of this chapter.

(J) If $\mathfrak{g}$ is a Lie algebra and $X \in \mathfrak{g}$, the operator ad X that maps Y to $[X, Y]$

is a linear transformation of $\mathfrak{g}$ into itself. It is easily verified that $X \to \text{ad } X$ is a representation of the Lie algebra $\mathfrak{g}$ with $\mathfrak{g}$ itself considered as the vector space of the representation. One need only check that $\text{ad}[X, Y] = [\text{ad } X, \text{ad } Y]$, and this is a simple consequence of the Jacobi identity. The representation $\text{ad } X$, called the *adjoint representation*, always provides a matrix representation of the algebra. If $\{E_i\}$ is a basis for $\mathfrak{g}$ then

$$\text{ad } E_i(E_j) = \sum_k C_{ij}^k E_k.$$

Therefore the matrix associated with the transformation $\text{ad } E_i$ is

$$(M_i)_{jk} = C_{ik}^j.$$

(Note the transposition of the indices j and k. Recall that if A is a linear transformation on a vector space V and that if $AE_i = \sum_j M_{ij} E_j$, where $\{E_i\}$ are the basis vectors of V, then the matrix of A is (M_{ji}).) As an exercise the reader may wish to check directly that $[M_a, M_b]_{jk} = \sum_r C_{ab}^r (M_r)_{jk}$. Another version of the adjoint representation is $(M_a)_{jk} = -C_{aj}^k$. (Right?)

The adjoint representation of $so(3)$, for example, is given by

$$(M_i)_{jk} = C_{ik}^j = \varepsilon_{ikj} = -\varepsilon_{ijk},$$

so the matrices L_1, L_2, and L_3 in (2.1) are in fact also the matrices of the adjoint representation.

(K) The *Killing form* of a Lie algebra is the symmetric bilinear form

$$K(X, Y) = \text{Tr}(\text{ad } X \text{ ad } Y).$$

Here $\text{ad } X$ and $\text{ad } Y$ are linear transformations on $\mathfrak{g}$.

If ρ is an automorphism of $\mathfrak{g}$ (that is ρ is a linear transformation of $\mathfrak{g}$ such that $\rho([X, Y]) = [\rho(X), \rho(Y)]$) then

$$K(\rho(X), \rho(Y)) = K(X, Y). \tag{2.7}$$

Moreover, K has the property

$$\begin{aligned} K([X, Y]), Z) &= K([Z, X], Y) \\ &= -K(Y, [X, Z]). \end{aligned} \tag{2.8}$$

If (E_i) form a basis for $\mathfrak{g}$ then

$$g_{ij} = K(E_i, E_j)$$

is called the *metric tensor* for $\mathfrak{g}$. In terms of the structure constants,

$$g_{ij} = \sum_{r,s} C_{is}^r C_{jr}^s.$$

For example, the metric tensor for $so(3)$ is $g_{ij} = -2\delta_{ij}$. For $sl(2, R)$ it is

$$\begin{pmatrix} 0 & 4 & 0 \\ 4 & 0 & 0 \\ 0 & 0 & 2 \end{pmatrix}.$$

As an exercise show that the metric tensor for $e(3)$ is

$$g = \begin{pmatrix} -4\mathbb{1} & 0 \\ 0 & 0 \end{pmatrix}$$

where $\mathbb{1}$ is the 3×3 identity matrix and the zeros denote the 3×3 zero matrix.

In Chapter 9 we shall prove *Cartan's Criterion*: a Lie algebra $\mathfrak{g}$ is semi-simple if and only if its Killing form is non-degenerate.

10. The Classical Lie Algebras

We conclude this chapter by deriving the Lie algebras of the classical groups. The procedures used here will be established rigorously in Chapter 3. We begin with $su(2)$ and generalize.

Typical elements of the Lie group $SU(2)$ are $U_1(\alpha) = e^{-i(\alpha/2)\sigma_1}$, $U_2(\beta) = e^{-i(\beta/2)\sigma_2}$, $U_3(\gamma) = e^{-i(\gamma/2)\sigma_3}$; the corresponding elements of the Lie algebra are $E_1 = \dfrac{-i\sigma_1}{2}$, $E_2 = \dfrac{-i\sigma_2}{2}$, $E_3 = \dfrac{-i\sigma_3}{2}$. Clearly the E_j can be obtained by differentiating the group elements at the identity: for example

$$E_1 = \frac{d}{d\alpha} U_1(\alpha)|_{\alpha=0}.$$

This method works for any linear Lie group $\mathfrak{G}$. Start with a curve $g(t)$ in the group that passes through the identity at $t = 0$. Then $X = \dfrac{d}{dt} g(t)|_{t=0}$ belongs to the Lie algebra and $\mathfrak{g}$ can be obtained by differentiating all such curves through the identity. Thus $\mathfrak{g}$ is the *tangent space to $\mathfrak{G}$ at the identity*. Conversely, if X belongs to the algebra then e^{tX} belongs to the group, just as we saw above for $SU(2)$.

In the case of linear Lie groups and algebras every group element near the identity can be written as e^{tX} where X belongs to the Lie algebra. We thus recover X as $\dfrac{d}{dt} e^{tX}|_{t=0}$. Near the identity $e^{tX} = I + tX \to O(t^2)$ and it is sometimes convenient in calculations to approximate elements near the identity by $I + tX$. For this reason elements of the algebra are sometimes called "elements of the infinitesimal group."

The Lie bracket of X_1 and X_2 is found by differentiating $e^{tX_1} e^{tX_2} e^{-tX_1} e^{-tX_2}$ at $t = 0$, and we do this in Chapter 3.

We now turn to the classical groups and their algebras. All of them are semi-simple with the exception of $gl(n, C)$. Curves through the identity in $GL(n, C)$ are of the form $I + tX + O(t^2)$, so $gl(n, C)$ consists of arbitrary $n \times n$

complex matrices. A useful basis for $gl(n, C)$, the *Weyl basis*, is the set of matrices $\{E_{ij}\}$, where E_{ij} has a 1 in the ijth entry and zeros everywhere else. Thus $gl(2, C)$ has the basis

$$
\begin{pmatrix} 1 & 0 \\ 0 & 0 \end{pmatrix}, \begin{pmatrix} 0 & 1 \\ 0 & 0 \end{pmatrix}, \begin{pmatrix} 0 & 0 \\ 1 & 0 \end{pmatrix}, \begin{pmatrix} 0 & 0 \\ 0 & 1 \end{pmatrix}.
$$

Now consider $SL(n, C)$. If $g(t)$ is a curve in $SL(n, C)$ we must have $\det g(t) \equiv 1$ for all t. When t is near zero, $g(t) = 1 + tX + O(t^2)$, so $\det g(t) = 1 + t\,\mathrm{Tr}\,X + O(t^2)$ and

$$
\frac{d}{dt} \det g(t)|_{t=0} = \mathrm{tr}\,X = 0
$$

Therefore $sl(n, C)$ is the algebra of $n \times n$ matrices with trace zero. For example, $sl(2, C)$ is spanned by $a_+ = \begin{pmatrix} 0 & 1 \\ 0 & 0 \end{pmatrix}$, $a_- = \begin{pmatrix} 0 & 0 \\ 1 & 0 \end{pmatrix}$, and $a_0 = \frac{1}{2}\begin{pmatrix} 1 & 0 \\ 0 & -1 \end{pmatrix}$.

It is easy to count the dimensions of these Lie algebras, that is, the dimension of each algebra considered as a vector space. Thus, $gl(n, C)$ has n^2 independent basis vectors (e.g. the Weyl basis above) so its dimension is n^2. The n^2 basis vectors obey one linear relation in $sl(n, C)$, so its dimension is $n^2 - 1$. We denote $sl(n, C)$ by A_{n-1} in Cartan's classification.

Clearly the dimensions of the real algebras $gl(n, R)$ and $sl(n, R)$ are also n^2 and $n^2 - 1$, respectively. (Had we considered the algebra $gl(n, C)$ *as an algebra over* $\mathbb{R}$ we would have needed $2n^2$ basis vectors, E_{ij} and $\sqrt{-1}E_{ij}$, for example.)

Elements $R(t)$ of the orthogonal group $SO(n)$ satisfy $R^t(t)R(t) = I$. Differentiating this at $t = 0$ we get $\dot{R}(0) + (\dot{R}(0))^t = 0$. The Lie algebra $so(n)$ is thus the algebra of skew symmetric matrices. A basis is the set $\{X_{ij} = E_{ij} - E_{ji}\}$, and there are $n(n-1)/2$ such matrices. When n is even, $n = 2m$, this algebra is D_m in Cartan's classification; when n is odd, $n = 2m + 1$, it is denoted by B_m.

Finally, elements $Q(t)$ in the symplectic groups $Sp(2n)$ satisfy $Q^t(t)JQ(t) = J$. Differentiating this at the identity gives $\dot{Q}^t(0)J + J\dot{Q}(0) \equiv X^tJ + JX = 0$. To analyze this it is easiest to set $X = \begin{pmatrix} A & B \\ C & D \end{pmatrix}$ with A, B, C, D being $n \times n$ matrices. The constraint $X^tJ + JX = 0$ implies that

$$
\begin{pmatrix} -C^t & A^t \\ -D^t & B^t \end{pmatrix} + \begin{pmatrix} C & D \\ -A & -B \end{pmatrix} = 0
$$

so that C and B are symmetric, $D = -A^t$, and A is arbitrary. So $sp(2n)$ is the algebra of matrices of the form $\begin{pmatrix} A & B \\ C & -A^t \end{pmatrix}$. There are n^2 basis elements needed for A, $\frac{1}{2}n(n+1)$ for B, and $\frac{1}{2}n(n+1)$ for C, giving $sp(2n)$ the dimension $n(2n + 1)$. This algebra is called C_n by Cartan.

We summarize these results in the following table:

The Classical Algebras

$\mathfrak{g}$	Constraint	Dimension
$sl(n) = A_{n-1}$	$\operatorname{Tr} X = 0$	$n^2 - 1$
$so(2n + 1) = B_n$	$X^t + X = 0$	$2n(2n + 1)$
$sp(2n) = C_n$	$\begin{pmatrix} A & B \\ C & -A^t \end{pmatrix}, B^t = B, C^t = C$	$n(2n + 1)$
$so(2n) = D_n$	$X^t + X = 0$	$n(2n - 1)$

These families include all but five of the complex semi-simple Lie algebras. The others, so-called *exceptional Lie algebras*, are denoted by E_6, E_7, E_8, F_4 and G_2 but are hardly discussed in this book.

CHAPTER 3

Lie Groups and Algebras: Matrix Approach

11. Exponentials and Logs

The Lie algebra $\mathfrak{g}$ of a Lie group $\mathfrak{G}$ is by definition the tangent space to $\mathfrak{G}$ (considered as an analytic manifold) at the identity. When $\mathfrak{G}$ is a matrix Lie algebra the elements of $\mathfrak{g}$ may be obtained by differentiating curves of matrices. For example, the curve in $SO(2)$ given by

$$
R(\theta) = \begin{pmatrix} \cos\theta & -\sin\theta \\ \sin\theta & \cos\theta \end{pmatrix}
$$

has as its tangent vector at the identity $\delta R = \dot{R}(0) = \begin{pmatrix} 0 & -1 \\ 1 & 0 \end{pmatrix}$.

Note that $R(\theta)$ is actually a one parameter group: that is $R(\theta + \mu) = R(\theta)R(\mu)$. Letting $L = \dot{R}(0)$ one finds without difficulty that

$$
\frac{dR}{d\theta} = LR(\theta).
$$

Thus the matrix $R(\theta)$ is a solution of a system of ordinary differential equations with constant coefficients. The solution of this equation is $R(\theta) = e^{\theta L} = I + \theta L + \frac{1}{2!}\theta^2 L^2 + \cdots$. It is easily seen that such a Taylor series always converges.

In general, a one-parameter group of matrices is a group such that $R(\theta + \mu) = R(\theta)R(\mu)$. The *infinitesimal generator* of such a group is $L = \delta R = \dot{R}(0)$, and $R(\theta) = e^{\theta L}$. In fact,

$$\frac{dR}{d\theta} = \lim_{h \to 0} \frac{R(\theta + h) - R(\theta)}{h}$$

$$= \lim_{h \to 0} \left(\frac{R(h) - I}{h} \right) R(\theta)$$

$$= \dot{R}(0) R(\theta) = LR(\theta).$$

The solution to this differential equation is $e^{\theta L}$.

If $g(\theta)$ is *any* curve in a Lie group for which $g(0) = I$, then $\dot{g}(0) = \lim_{\theta \to 0} \frac{g(\theta) - I}{\theta}$ is the tangent vector at the identity. The curve $g(\theta)$ need not be a one-parameter group. The set of all such tangent vectors spans the Lie algebra. This was the procedure we used to calculate the Lie algebras of the classical groups at the end of Chapter 2; but the same procedure works for all matrix groups. For example, the group of all upper triangular matrices

$$g(\alpha, \beta, \gamma) = \begin{pmatrix} 1 & \alpha & \beta \\ 0 & 1 & \gamma \\ 0 & 0 & 1 \end{pmatrix}$$

is a three parameter group whose Lie algebra is generated by the three matrices $\frac{\partial g}{\partial \alpha}, \frac{\partial g}{\partial \beta},$ and $\frac{\partial g}{\partial \gamma}$ evaluated at $\alpha = \beta = \gamma = 0$.

Matrix representations of Lie groups and algebras are particularly simple to work with, and we shall focus on these in the present chapter. We begin by proving:

Theorem 3.1. *Let $\mathfrak{G}$ be a matrix Lie group and let $\mathfrak{g}$ be the set of tangent vectors to all curves in $\mathfrak{G}$ at the identity. Then $\mathfrak{g}$ is a Lie algebra of matrices with the commutator as the Lie product.*

PROOF. Let us show first that $\mathfrak{g}$ is a linear vector space. Thus, if $A, B \in \mathfrak{g}$ we must show that $A + B$ and λA belong to $\mathfrak{g}$, where λ is any scalar. Let $U(\theta)$ and $V(\theta)$ be curves in $\mathfrak{G}$ with $U(0) = V(0) = I$, $\delta U = A$, $\delta V = B$. Then $U(\theta) V(\theta)$ is a curve in $\mathfrak{G}$ and $\delta(UV) = (\delta U) V(0) + U(0) \delta V = A + B$, so $A + B \in \mathfrak{g}$. Furthermore, $U(\lambda \theta) \in \mathfrak{G}$ and $\delta U(\lambda \theta) = \lambda \delta U = \lambda A \in \mathfrak{g}$.

Next we show that $[A, B] \in \mathfrak{g}$ as well. To this end we must suppose that the curves $U(\theta)$ and $V(\theta)$ are smooth, say at least of class C^3, so that we may expand U and V in a Taylor series:

$$U(\theta) = I + \theta A + \frac{\theta^2}{2} \ddot{U}(0) + \cdots$$

$$U^{-1}(\theta) = I - \theta A - \frac{\theta^2}{2}(\ddot{U}(0) - 2A^2)) + \cdots$$

and similarly for $V(\theta)$ and $V^{-1}(\theta)$. Now define $h(\tau) = U(\theta)V(\theta)U^{-1}(\theta)V^{-1}(\theta)$, where $\tau = \theta^2$. By expanding $h(\tau)$ in a Taylor series we find that $h(\tau) = I + \tau[A, B] + O(\theta^3)$, and therefore $\delta h = [A, B]$. Since $h(\tau)$ lies in $\mathfrak{G}$ we may conclude that $[A, B] \in \mathfrak{g}$, and Theorem 3.1 is proved. $\square$

The converse to Theorem 3.1 is also true: given a matrix Lie algebra there is a Lie group with that algebra as its tangent space at the identity. In order to construct the group from the algebra we need to develop some more machinery.

Let $M_n(C)$ be the algebra of $n \times n$ matrices with complex entries. $M_n(C)$ is a *Banach space* with the norm

$$\|A\| = \sup_{\|v\|} \frac{\|Av\|}{\|v\|}$$

where $v \in C^n$ and $\|v\|^2 = \sum v_i \bar{v}_i$. That $M_n(C)$ is a Banach space means that every Cauchy sequence in $M_n(C)$ converges to an element in $M_n(C)$; thus, if $\{A_k\}$ is a sequence such that $\|A_k - A_j\| \to 0$ as $k, j \to \infty$ then there exists $A \in M_n(C)$ such that $\|A_k - A\| \to 0$. Moreover, if $\|A_k - A\| \to 0$ then each entry of A_k must converge to the corresponding entry of A. $M_n(C)$ with the commutator $[A, B] = AB - BA$ as its Lie bracket is the Lie algebra $\mathfrak{gl}(n, C)$.

Lemma 3.2. *Let $f(z) = \sum a_k z^k$ be a convergent power series with radius of convergence r. Then the series $f(A) = \sum a_k A^k$ converges in the norm of $M_n(C)$ for all A such that $\|A\| < r$.*

The proof of this simple lemma is left to the reader. An immediate consequence is that the matrix e^{zA} is defined for any matrix A and any complex number z. A second consequence is that $\log B$ is uniquely defined for all B such that $\|I - B\| < 1$. This follows from the fact that

$$\log z = \log[1 + (z - 1)] = \sum_{n=1}^{\infty} (-1)^{n+1} \frac{(z - 1)^n}{n}$$

converges for $|z - 1| < 1$. Furthermore, since $e^{\log z} = z$ is an identity in $|z - 1| < 1$, $e^{\log B} = B$ is an identity on $\|B - I\| < 1$. We have proved

Lemma 3.3. *The mapping $B \to \log B$ is an analytic mapping of the neighborhood $U = \{B | \|B - I\| < 1\}$ of the identity in $GL(n, C)$ into the algebra $gl(n, C)$.*

The relationship between $gl(n, C)$ and a neighborhood of the identity in $GL(n, C)$ given in Lemma 3.3 carries over to matrix Lie groups and algebras in general. If $\mathfrak{g}$ is a Lie algebra and $X_1 \ldots X_n$ form a basis for $\mathfrak{g}$, then $e^{\alpha_1 X_1}, \ldots, e^{\alpha_n X_n}$ generate the connected component of the corresponding Lie group. Similarly, if B is an element of the Lie group and $\|B - 1\| < 1$, then $\log B$ lies in the algebra. The proof of this fact requires the use of the Campbell–Baker–Hausdorff theorem, to which we now turn.

Recall that the adjoint representation $A \to \text{ad } A$ is defined by $\text{ad } A(B) = [A, B]$, for A and B in a Lie algebra $\mathfrak{g}$. Consider $\rho_\theta(B) = e^{\theta A} B e^{-\theta A}$. We have

$$\frac{d}{d\theta} \rho_\theta(B) = [A, \rho_\theta(B)] = \text{ad } A(\rho_\theta(B)).$$

The solution of this differential equation is

$$\rho_\theta(B) = e^{\theta \, \text{ad } A}(B)$$

$$= I + \theta[A, B] + \frac{\theta^2}{2!}[A, [A, B]] + \cdots.$$

Therefore, if $A, B \in \mathfrak{g}$ so does $e^{\theta \, \text{ad } A} B$.

The solution of the inhomogeneous equation

$$\frac{du}{ds} = Au + w$$

where u and w assume values in a vector space V and A is a linear transformation on V, is given by

$$u(s) = e^{sA} u(0) + \int_0^s e^{(s-t)A} w \, dt.$$

If w is a constant vector the integration may be carried out to give

$$u(s) = e^{sA} u(0) + f(s, A)w \qquad (3.1)$$

where $f(s, z) = (e^{sz} - 1)/z$. Since $f(s, z)$ is an entire function of z, $f(s, A)$ is defined for any transformation A.

This result may be used to prove the following:

Lemma 3.4. *Let $A(t)$ be any matrix valued function of t. Then*

$$e^{A(t)} \frac{d}{dt} e^{-A(t)} = -f(\text{ad } A(t))\dot{A}$$

where $f(z) = (e^z - 1)/z$.

PROOF. Let $B(s, t) = e^{sA(t)} \dfrac{d}{dt} e^{-sA(t)}$. Then

$$\frac{\partial B}{\partial s} = [A, B] - \dot{A}(t), \qquad B(0, t) = 0.$$

The solution of this ordinary differential equation in s is given by (3.1) with A replaced by Ad A. The result follows on setting $s = 1$. $\qquad\qquad\square$

Theorem 3.5 (Campbell–Baker–Hausdorff). *For A, B sufficiently close to the origin in $M_n(C)$ the matrix $C = \ln e^A e^B$ is uniquely defined, and*

$$C = B + \int_0^1 g(e^{t\,\mathrm{ad}\,A} e^{\mathrm{ad}\,B}) A\, dt \tag{3.2}$$

where $g(z) = \ln z/(z-1)$.

PROOF. Let $C(t) = \ln e^{tA} e^B$. Then

$$e^{C(t)} \frac{d}{dt} e^{-C(t)} = -A$$

so, by Lemma 3.4,

$$A = f(\mathrm{ad}\, C(t))(\dot{C}). \tag{3.3}$$

Now $e^{\mathrm{ad}\, C(t)} h = e^{C(t)} h e^{-C(t)} = e^{tA} e^B h e^{-B} e^{-tA} = e^{t\,\mathrm{ad}\,A} e^{\mathrm{ad}\,B}(h)$; and therefore $e^{\mathrm{ad}\, C(t)} = e^{t\,\mathrm{ad}\,A} e^{\mathrm{ad}\,B}$. For A and B sufficiently small $\ln e^{t\,\mathrm{ad}\,A} e^{\mathrm{ad}\,B}$ is defined, and

$$\mathrm{ad}\, C(t) = \ln e^{t\,\mathrm{ad}\,A} e^{\mathrm{ad}\,B}$$

From the identity $f(\ln z)g(z) = 1$ we see that

$$f(\mathrm{ad}\, C(t)) = f(\ln e^{t\,\mathrm{ad}\,A} e^{\mathrm{ad}\,B})$$
$$= g(e^{t\,\mathrm{ad}\,A} e^{\mathrm{ad}\,B})^{-1}$$

and therefore, from (3.3)

$$\dot{C}(t) = g(e^{t\,\mathrm{ad}\,A} e^{\mathrm{ad}\,B}) A.$$

Integrating this from 0 to 1, we get (3.2).

By carrying out the integration termwise we obtain the first few terms:

$$C = A + B + \frac{1}{2}[A,B] + \frac{1}{12}([A,[A,B]] - [B,[B,A]]) + \cdots. \tag{3.4}$$

Since A and B belong to $\mathfrak{g}$ so does C. $\qquad\square$

We are now ready to prove that every linear Lie algebra generates a Lie group. The theorem is purely a local one, and so we must begin by defining a local linear Lie group.

Definition 3.6. *A local linear Lie group is a set of matrices parameterized by a matrix valued function $g(\theta)$, depending analytically on θ as θ varies over a domain $U \subset \mathbb{R}^n$ (or $\mathbb{C}^n$), with the following properties*

(i) *U contains the origin and $g(0) = I$.*
(ii) *g is one-to-one over U.*
(iii) *There exist $V \subset U$ and an analytic mapping h from $V \times V$ to U such that for $\theta, \psi \in V$, $g(\theta)g(\psi) = g(h(\theta, \psi))$.*
(iv) *For $\theta \in V$ there is a unique ψ in V such that $g(\theta)g(\psi) = I$, and the mapping $\theta \to \psi$ is analytic.*

Theorem 3.7. *Let g be a linear Lie algebra generated by $L_1, \ldots, L_n$. Then $g(\theta) = e^{\theta_1 L_1 + \cdots + \theta_n L_n}$ is a local linear Lie group for sufficiently small $|\theta|$.*

PROOF. For $|\theta|$ sufficiently small $\|g(\theta) - I\| < 1$ and so $\log g(\theta)$ is uniquely defined; hence $g(\theta)$ is one-to-one for sufficiently small $|\theta|$. The Campbell–Baker–Hausdorff theorem shows that for sufficiently small θ and ψ there is a unique matrix $C(\theta, \psi)$ such that $e^{C(\theta, \psi)} = g(\theta)g(\psi)$. Furthermore, formula (3.2) makes it clear that C depends analytically on θ and ψ. We may write

$$C(\theta, \psi) = \sum_{j=1}^{n} h^j(\theta, \psi) L_j.$$

The h^j, which depend analytically on θ and ψ, form the components of the composition function h described in part (iii) of Definition (3.6). The inverse of part (iv) is obtained simply by taking $\psi = -\theta$, since $(e^L)^{-1} = e^{-L}$ for any matrix L. This concludes the proof.   □

Global questions are more subtle. For example, let $\mathfrak{G}$ be a linear Lie group and $\mathfrak{g}$ its Lie algebra. Then every element of $\mathfrak{G}$ need not be the exponential of an element of the algebra, even if $\mathfrak{G}$ is connected. (see exercise 4). For another example, although the Lie algebras of $SU(2)$ and $SO(3)$ are isomorphic, $SU(2)$ and $SO(3)$ are different as Lie groups, since $SU(2)$ is simply connected, while $SO(3)$ is not.

For linear Lie algebras the proof that their exponentials generate a Lie group is relatively simple. The fact that every Lie algebra over $\mathbb{R}$ or $\mathbb{C}$ is the Lie algebra of a Lie group is a much deeper result and requires the integration of overdetermined systems of differential equations (Chapter 8). The integration theory in which the group is constructed from the algebra is due to Lie himself and was a real *tour de force* in the development of the subject.

Ado's theorem on the other hand, states that every Lie algebra is isomorphic to a matrix Lie algebra; and this fact obviates Lie's original method. Given any Lie algebra we may construct a representation of it by a linear Lie algebra and so every Lie group is locally isomorphic to a matrix group. This situation, however, does not in general hold in the large (see G. Birkhoff).

A *canonical* coordinate system of the first kind is a system $\theta^1, \ldots, \theta^n$ such that the curves $\theta^i(t) = ta^i$ (a^i constant) are all one parameter subgroups (when lifted to the group). In our proof of Theorem 3.7 we chose the coordinates $\theta^1, \ldots, \theta^n$ to be such coordinates. We might also have taken $g(\theta) = e^{\theta_1 L} \ldots e^{\theta_n L_n}$, but these coordinates are not canonical. In the case of the rotation group $SO(3)$ there are two well-known coordinate systems. The *Euler angles* are those for which a general element is represented by

$$R(\alpha, \beta, \gamma) = R_z(\alpha) R_y(\beta) R_z(\gamma)$$

where $R_z(\gamma)$ is a rotation through angle γ about the z-axis, etc. These are not canonical coordinates. The other parameterization that we described in Chapter 1 was the representation of a rotation by a point p in the ball of radius π, with p representing a rotation through an angle $|p|$ about the axis $p/|p|$. These coordinates do form a canonical coordinate system of the first kind, since the ray $\theta\,\hat{n}$ represents a rotation through angle θ about the fixed axis $\hat{n}$. In particular, in the spinor representation $e^{-(\theta/2)\hat{n}\cdot\vec{\sigma}}$, θ and $\hat{n}$ are canonical coordinates of the first kind.

12. Automorphisms and Derivations

We now take a closer look at the relationship between a group and its algebra.

An algebra $\mathscr{A}$ is a linear vector space on which there is also a process of multiplication $(A, B) \to A \cdot B$ which is distributive with respect to addition. A Lie algebra is an algebra in this sense with the Lie product as the operation of multiplication. An *automorphism* ρ of an algebra is a one to one mapping of $\mathscr{A}$ onto itself which preserves both multiplication and addition:

$$\rho(\alpha A + \beta B) = \alpha\rho(A) + \beta\rho(B),$$

$$\rho(A \cdot B) = \rho(A) \cdot \rho(B).$$

Here are some examples:

(i) $M_n(C)$ is an algebra with matrix addition and multiplication as the operations. If P is a non-singular matrix then $\rho_p(A) = PAP^{-1}$ is an automorphism of $M_n(C)$.

(ii) $C^\infty(R^n)$ is an algebra with multiplication defined by $(fg)(x) = f(x)g(x)$. If ψ is a smooth transformation of R^n the operation $\rho_\psi f = f \circ \psi$ (i.e., $\rho_\psi f(x) = f(\psi(x))$ is an automorphism of $C^\infty(R^n)$).

(iii) If $\mathfrak{g}$ is a matrix Lie algebra and $A \in \mathfrak{g}$, then $\rho_A(B) = e^{\theta A}Be^{-\theta A} = e^{\theta\,\mathrm{ad}\,A}(B)$ is a one-parameter group of automorphisms of $\mathfrak{g}$ as we saw in §11.

A *derivation D* on an algebra $\mathscr{A}$ is a linear operation such that for $f, g \in \mathscr{A}$ and scalars α and β,

$$D(\alpha f + \beta g) = \alpha Df + \beta Dg,$$

$$D(fg) = (Df)g + f(Dg).$$

For example, the first order partial differential operator

$$X = \sum_{i=1}^{n} X^i \frac{\partial}{\partial x^i},$$

where $X^i \in C^\infty(R^n)$, is a derivation on $C^\infty(R^n)$. (Later we shall see that all derivations of $C^\infty(R^n)$ are of this form.) As another example, if A belongs to a Lie algebra $\mathfrak{g}$ then ad A is a derivation on $\mathfrak{g}$; again this follows from the Jacobi identity.

There is a very close connection between derivations and one parameter groups of automorphisms of an algebra. Namely, if ρ_θ is a one-parameter group of automorphisms of an algebra $\mathcal{A}$, then its infinitesimal generator is a derivation. In fact, if $Df = \delta\rho_\theta f$, then from

$$\rho_\theta(fg) = (\rho_\theta f)(\rho_\theta g)$$

we get

$$Dfg = \delta\rho_\theta(fg) = (\delta\rho_\theta f)g + f\delta\rho_\theta g$$
$$= (Df)g + f(Dg),$$

so D is a derivation.

To take a simple example, let ρ_θ be the one parameter group of automorphisms of $C^\infty(R)$ given by $(\rho_\theta f)(x) = f(x + \theta)$. The generator of this group is the derivative d/dx; in fact,

$$\delta(\rho_\theta f)(x) = \frac{d}{d\theta}f(x + \theta)|_{\theta=0} = \frac{df}{dx}.$$

As another example, the infinitesimal generator of the group of automorphisms of a Lie algebra $\mathfrak{g}$ given by $\rho_\theta(B) = e^{\theta A}Be^{-\theta A}$ is, as we have seen, ad A. We state this example as a Theorem, for emphasis.

Theorem 3.8. *The adjoint representation of a linear Lie algebra $\mathfrak{g}$, given by* ad A, *is a derivation of the algebra. It is the infinitesimal generator of the one parameter group of automorphisms defined by* $\rho_\theta(B) = e^{\theta A}Be^{-\theta A} = e^{\theta\,\mathrm{ad}\,A}(B)$.

EXERCISES

1. Compute $\exp t\begin{pmatrix} \lambda & 1 \\ 0 & \lambda \end{pmatrix}$.

2. Let $D = d/dx$, let f be analytic in x, and compute $e^{\theta D}f$ by expanding $e^{\theta D}$ in a Taylor series.

3. Prove that $\det e^A = e^{\mathrm{Tr}\,A}$. If A is diagonal (or diagonalizeable) it is immediate. If A is not diagonal, prove the result by putting A in Jordan canonical form. Another method is this: Let $\psi(t) = \det e^{tA}$; then $\psi(t + s) = \psi(t)\psi(s)$. What is $\psi(0)$?

4. Let $A \in sl(2, R)$ and let $\sigma = \sqrt{\det A}$, with $\sigma = i\sqrt{|\det A|}$ if $\det A < 0$. Show that

$$e^A = \cos\sigma I + \left(\frac{\sin\sigma}{\sigma}\right)A.$$

Show that $\begin{pmatrix} \lambda & 0 \\ 0 & \lambda^{-1} \end{pmatrix} \in SL(2, R)$, but cannot be represented in the form e^A for any $A \in sl(2, R)$ if $\lambda < 0$ and $\lambda \neq -1$. Thus, even though $SL(2, R)$ is connected it cannot be covered by the exponential mapping of its Lie algebra.

5. If $\mathfrak{s}$ is a Lie subalgebra of $\mathfrak{g}$ then the Lie group generated by $\mathfrak{s}$ is a subgroup of that generated by $\mathfrak{g}$; that subgroup is normal if $\mathfrak{s}$ is an ideal.

6. Let $X = (X^1, \ldots X^n)$ be a C^∞ vector field on R^n and let $\psi(t, x)$ be the flow generated by the solution set of the differential equations $\dot{x}^i = X^i(x^1 \ldots x^n)$. Show $\psi(t, \psi(s, x)) = \psi(t + s, x)$. Define $\rho_t f = f \circ \psi_t$. Show that ρ_t is a one parameter group of automorphisms of $C^\infty(R^n)$ and find its infinitesimal generator.

7. Let $\dot{x} = Ax$, $x = R^n$ be a linear system of differential equations in R^n. Show the flow generated by this system is given by $\rho_t(x) = e^{tA}x$. Let $A = \begin{pmatrix} 0 & -1 \\ 1 & 0 \end{pmatrix}$ and use the group property $\rho_{t+s} = \rho_t \rho_s$ to derive the addition formulae for the sine and cosine. Let $z = y/x$ and find a differential equation for z. Denote this flow by $\phi_t(z)$. Use the homomorphic property $\phi_{t+s} = \phi_t \circ \phi_s$ to prove the addition formula for the tangent function.

CHAPTER 4

Applications to Physics and Vice Versa

13. Poisson Brackets and Quantization

Lie groups and their algebras arise most often in physics as symmetry groups of dynamical systems. These symmetries are intimately associated with conservation laws. For example, if a physical system is invariant under translations then its linear momentum is conserved; while rotational invariance of a system implies conservation of angular momentum. In modern physics the symmetry groups are not only the geometrical symmetries of space-time, but also new symmetries associated with "internal" degrees of freedom of particles and fields. These symmetries lead to the conservation of more exotic quantities such as *isospin, strangeness, charm,* etc.

The state of a classical dynamical system is described in Hamiltonian mechanics by giving N coordinates $q_1, \ldots, q_n$ and N momenta $p_1, \ldots, p_n$. The $2N$ variables $\{q_1, \ldots, p_n\}$ are referred to collectively as canonical variables of the system. Other physically important quantities such as energy and momentum are functions $F = F(q, p)$ of the canonical variables. These functions, called observables, form an infinite dimensional Lie algebra with respect to the Poisson bracket

$$\{F, G\} = \sum_{i=1}^{N} \left(\frac{\partial F}{\partial q_i} \frac{\partial G}{\partial p_i} - \frac{\partial F}{\partial p_i} \frac{\partial G}{\partial q_i} \right).$$

The equations of motion are $\dot{q}_i = \dfrac{\partial H}{\partial p_i}$ and $\dot{p}_i = -\dfrac{\partial H}{\partial q_i}$, where H, the Hamiltonian of the system, is the total energy. These equations, called Hamilton's equations, may be written in terms of Poisson brackets as

$$\dot{q}_i = \{q_i, H\}, \qquad \dot{p}_i = \{p_i, H\}.$$

More generally, the time evolution of an observable F is given by

$$\dot{F} = \{F, H\}. \tag{4.1}$$

A simple example will help fix these ideas and illustrate the relationship between symmetries and conservation laws. Consider a system consisting of two particles on a line. We label the canonical variables $\{q_1, q_2, p_1, p_2\}$ with q_i representing the position of the ith particle and p_i its momentum. Translation of the coordinate system by an amount x means that the canonical variables in the new system are $\{q'_1, q'_2, p'_1, p'_2\} = \{q_1 - x, q_2 - x, p_1, p_2\}$. This induces an automorphism on the observable $F \rightarrow F'$ given by $F'(q'_1, q'_2, p'_1, p'_2) = F(q_1, q_2, p_1, p_2)$. (That is, the new function evaluated in the new coordinates is numerically equal to the old function evaluated in the old coordinates.) The infinitesimal generator of this one-parameter transformation group is

$$\frac{dF'}{dx}(q'_1, q'_2, p'_1, p'_2)|_{x=0} = \frac{d}{dx} F(q'_1 + x, q'_2 + x, p'_1, p'_2)|_{x=0}$$

$$= \frac{\partial F}{\partial q_1} + \frac{\partial F}{\partial q_2}.$$

Now $\dfrac{\partial F}{\partial q_i} = \{F, p_i\}$ so the infinitesimal generator may be expressed as $\{F, p_1 + p_2\}$. Then, if the Hamiltonian H is invariant under translations, we have $\{H, p_1 + p_2\} = 0$. By (4.1) the quantity $p_1 + p_2$ that generates the translation in space is therefore constant in time:

$$\frac{d}{dt}(p_1 + p_2) = \{p_1 + p_2, H\} = 0.$$

In general, if G is a function of the canonical variables such that $\{G, H\} = 0$ where H is the Hamiltonian, then G generates a symmetry of the system found by solving the equations

$$\frac{dq_i}{ds} = \{q_i, G\} = \frac{\partial G}{\partial p_i}$$

$$\frac{dp_i}{ds} = \{p_i, G\} = -\frac{\partial G}{\partial q_i};$$

for if $\{q(s), p(s)\}$ is the flow generated by these equations we have

$$\frac{dH}{ds} = \sum_i \left(\frac{\partial H}{\partial q_i} \frac{dq_i}{ds} + \frac{\partial H}{\partial p_i} \frac{dp_i}{ds} \right)$$

$$= \sum_i \left(\frac{\partial H}{\partial q_i} \frac{\partial G}{\partial p_i} - \frac{\partial H}{\partial p_i} \frac{\partial G}{\partial q_i} \right)$$

$$= \{H, G\} = 0.$$

Then H is invariant under the flow: $H(q(s), p(s)) = H(q_0, p_0)$.

The reader may easily check that the symmetries of the Hamiltonian—that is, the set of functions G such that $\{H, G\} = 0$—form a Lie subalgebra of the Lie algebra of all observables.

We have just seen that in the classical description of a system the observables are functions of the positions and momenta, and these observables form a Lie algebra whose Lie product is the Poisson bracket. A quantum mechanical description of the same system is obtained by finding an algebra of Hermitian operators on a Hilbert space with the Lie product given by the commutator. This is to be done in such a way that if A and B are operators corresponding to the classical functions a and b, then $[A, B]$ is an operator corresponding to the classical function $i\hbar\{a, b\}$ ($\hbar = h/2\pi$ where h is Planck's constant). In particular, since the classical canonical variables q_r, p_s satisfy $\{q_r, q_s\} = 0, \{p_r, p_s\} = 0, \{q_r, p_s\} = \delta_{rs}$ the corresponding quantum mechanical operators must satisfy the commutation relations

$$[Q_r, Q_s] = 0, \qquad [P_r, P_s] = 0, \qquad [Q_r, P_s] = i\hbar\delta_{rs}\mathbb{1}.$$

One realization of such an algebra is given by operators acting on functions of the coordinates $q_1, \ldots, q_N$:

$$Q_r\Psi(q_1, \ldots, q_n) = q_r\Psi(q_1, \ldots, q_n)$$

$$P_s\Psi(q_1, \ldots, q_n) = -i\hbar\frac{\partial\Psi}{\partial q_s}(q_1, \ldots, q_n).$$

(N.B. The quantization prescription that associates an operator B with a function b is ambiguous since several operators may correspond to the same classical quantity. For example, the operators $QP^2 + P^2Q$ and PQP might both correspond to the classical quantity qp^2.)

In the case of one variable we obtain the Heisenberg algebra $\{Q, P, E\}$ where $E = i\hbar\mathbb{1}$; with n variables we get the n-fold direct sum. A useful form of the Heisenberg algebra is found by setting $a = \dfrac{P - iQ}{\sqrt{2}}$, $a^* = \dfrac{P + iQ}{\sqrt{2}}$, so that

$[a, a] = [a^*, a^*] = 0$ and $[a, a^*] = \mathbb{1}$.

The Heisenberg algebra lies at the heart of quantum mechanics. We will illustrate its use in §15 by studying a simple quantum mechanical problem: the harmonic oscillator.

EXERCISES

1. A system of N bodies in space has the Hamiltonian

$$H = \sum_{i=1}^{N}\sum_{\alpha=1}^{3}\frac{p_{i\alpha}^2}{2m_i} + \sum_{i>j}V(|\vec{x}_i - \vec{x}_j|).$$

$p_{i\alpha}$ is the αth component of momentum of the ith particle and m_i is its mass; $x_i = (x_{i1}, x_{i2}, x_{i3})$. Show that H is invariant under translations and rotations, and find the corresponding conservation laws.

2. A transformation $\tilde{x}_j = X_j(x, p)$, $\tilde{p}_j = P_j(x, p)$ is *symplectic* if it preserves the Poisson bracket $\{\,,\,\}$. That is

$$\sum_i \frac{\partial F}{\partial \tilde{x}_i} \frac{\partial G}{\partial \tilde{p}_i} - \frac{\partial F}{\partial \tilde{p}_i} \frac{\partial G}{\partial \tilde{x}_i} = \sum_i \frac{\partial F}{\partial x_i} \frac{\partial G}{\partial p_i} - \frac{\partial F}{\partial p_i} \frac{\partial G}{\partial x_i}.$$

Prove that X, P is a symplectic transformation iff

$$\{X_i, X_j\} = 0 \qquad \{X_i, P_j\} = \delta_{ij} \qquad \{P_i, P_j\} = 0.$$

3. Prove that the flow generated by a Hamiltonian vector field is a one-parameter group of symplectic transformations.

4. Let X_j, P_j be the symplectic transformation given above. Let

$$A = \begin{pmatrix} \dfrac{\partial X_i}{\partial x_j} & \dfrac{\partial X_i}{\partial p_k} \\[2ex] \dfrac{\partial P_l}{\partial x_j} & \dfrac{\partial P_l}{\partial p_k} \end{pmatrix}.$$

Show that A^t is a symplectic matrix.

14. Motion of a Rigid Body; Euler's Equations

Consider the problem of describing the motion of a rigid body. Let all points be referred to a fixed "inertial" coordinate system. Then the instantaneous configuration of the body is prescribed by giving the coordinates of its center of mass and the "attitude" of the body in space. The latter is uniquely given by an orthogonal transformation $O(t)$ which relates a fixed inertial frame $\vec{e}_1$, $\vec{e}_2$, $\vec{e}_3$ to a moving frame $\vec{E}_1(t)$, $\vec{E}_2(t)$, $\vec{E}_3(t)$ attached to the body. The configuration space of the body may therefore be identified with the group of rigid motions in $\mathbb{R}^3$.

More generally, problems of describing the motion of a dynamical system in a moving coordinate system arise, for example, in meteorology or oceanography, where the earth's rotation introduces centrifugal and Coriolis forces into the equations of motion. Let $\vec{e}_1$, $\vec{e}_2$, and $\vec{e}_3$ be a fixed "inertial" frame (of

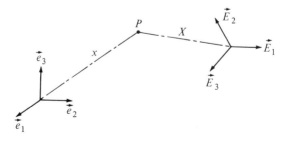

Figure 4.1. Coordinates of P relative to two frames.

course, such frames do not exist in relativity theory, but we are operating in the realm of Newtonian mechanics here) and let $\vec{E}_1(t)$, $\vec{E}_2(t)$, $\vec{E}_3(t)$ be an orthonormal frame moving in space. Let x^j and X^j denote the coordinates of a point P relative to the two frames (see Figure 4.1). Assuming the transformation between the two frames to be a rigid motion, the coordinates are related by

$$x^j = O_k^j X^k + r^j$$

where $O(t)$ is an orthogonal matrix. We shall call the x^j "rest-frame" or "inertial" coordinates and the X^k moving coordinates.

Newton's equations of motion of a particle in inertial coordinates are

$$\frac{m d^2 x^j}{dt^2} = f^j(x, v)$$

where m is the mass of the particle, v is the velocity and f^j are the components of the force. We want to write these equations in a form which is correct in any moving frame. Such a form of the equations is said to be *gauge invariant*.

We drop the translation r (it is the simpler of the two effects) and find the gauge invariant form for rotations alone. Thus we assume that $x^j = O_k^j X^k$, or, more simply, $x = OX$, where $O(t)$ is an orthogonal matrix. The relationship between the relative velocity $\dot{X}$ and the absolute velocity $\dot{x}$ is easily found. We have

$$\dot{x} = O\dot{X} + \dot{O}X$$
$$= O(\dot{X} + \Omega X)$$

where $\Omega = O^t\dot{O}$. It is easily seen that $\Omega \in so(3)$. We write this relationship in the form

$$v = \dot{x} = OV$$

where $V = D_t X = \dot{X} + \Omega X$ is the "relative velocity." In this way we have

$$x = OX, \qquad v = OV.$$

The operator $D_t = \dfrac{\partial}{\partial t} + \Omega$ is called the total derivative in mechanics. It is an example of a connexion; the connexion term Ω adjusts for the rotation of the moving coordinate frame. Another way to express this is to write

$$D_t X = O^t \frac{d}{dt} OX.$$

In other words, to find the relative velocity, we transform to the inertial frame, differentiate with respect to time, and transform back.

Newton's equations of motion in an arbitrary frame are then

$$m D_t^2 X = F(X, V) \tag{4.2}$$

where $OF(X, V) = f(OX, OV)$. In fact

$$mD_t^2 X = m(O^t \frac{d}{dt} O)(O^t \frac{d}{dt} O) X$$

$$= mO^t \ddot{x} = O^t f(x, v) = O^t f(OX, OV).$$

The form (4.2) of Newton's equations are gauge invariant: they are correct in any rotating frame (gauge) provided one takes $V = D_t X$.

Let us calculate $D_t^2 X$:

$$D_t^2 X = D_t(\dot{X} + \Omega X)$$

$$= \ddot{X} + \dot{\Omega} X + 2\Omega \dot{X} + \Omega^2 X.$$

The term $2m\Omega\dot{X}$ is the Coriolis force and $m\Omega^2 X$ is the centrifugal force; $m\dot{\Omega}X$ is inertial force due to the rotation of the coordinate system.

Let us now find the equations for the angular momentum. As a consequence we obtain Euler's equations of motion of a rigid body. These are traditionally written in terms of the angular momentum vector. In three dimensions there is a simple relationship between vectors and skew-symmetric second order tensors, namely

$$\omega = (\omega_1, \omega_2, \omega_3) \rightarrow \Omega = \begin{pmatrix} 0 & -\omega_3 & \omega_2 \\ \omega_3 & 0 & -\omega_1 \\ -\omega_2 & \omega_1 & 0 \end{pmatrix}.$$

In fact, we have already seen such a relationship, *viz.* we may map vectors $\vec{x}$ in $\mathbb{R}^3$ into elements $\mathfrak{x}$ in $so(3)$ via $\mathfrak{x} = \sum_{j=1}^3 x^j L_j$, where L_i are matrices (2.1) which span $so(3)$.

The reader may easily verify that $\vec{x} \rightarrow \mathfrak{x}$ is an isomorphism from the Lie algebra $\{\mathbb{R}^3, \times\}$ to $so(3)$; that is $\vec{u} \times \vec{v} \rightarrow [\mathfrak{u}, \mathfrak{v}]$. Moreover, $\mathfrak{x} = O \mathfrak{X} O^t$. In this representation $D_t \mathfrak{X} = \dot{\mathfrak{X}} + [\Omega, \mathfrak{X}]$. In fact,

$$v = \dot{\mathfrak{x}} = \frac{d}{dt} O \mathfrak{X} O^t$$

$$= \dot{O} \mathfrak{X} O^t + O \dot{\mathfrak{X}} O^t + O \mathfrak{X} \dot{O}^t$$

$$= O(\dot{\mathfrak{X}} + [\Omega, \mathfrak{X}]) O^t.$$

As before we take $\mathfrak{V} = D_t \mathfrak{X}$.

The angular momentum vector $\vec{l}$ is given by $\vec{l} = \vec{x} \times m\vec{v}$; in this representation, therefore,

$$\mathfrak{l} = m[\mathfrak{x}, \mathfrak{v}] \quad \text{and} \quad \mathfrak{L} = O^t \mathfrak{l} O = m[\mathfrak{X}, \mathfrak{V}].$$

Newton's equations of motion are $(d/dt)\vec{l} = m(d\vec{x}/dt) \times \vec{v} + m\vec{x} \times (d\vec{v}/dt) = m\vec{x} \times (d\vec{v}/dt) = \vec{x} \times \vec{f} = \vec{\tau}$ where $\vec{\tau}$ is the torque. If $\vec{\tau} = 0$ then $d\vec{l}/dt = 0$. Newton's equations of motion are therefore

$$D_t \mathfrak{L} = \dot{\mathfrak{L}} + [\Omega, \mathfrak{L}] = 0. \tag{4.3}$$

Now $\mathfrak{B} = D_t\mathfrak{X} = \dot{\mathfrak{X}} + [\Omega, \mathfrak{X}]$. If we choose a frame which rotates with the body, then $\dot{\mathfrak{X}} = 0$, and

$$\mathfrak{L} = m[\mathfrak{X}, [\Omega, \mathfrak{X}]] = -m(\mathrm{ad}\,\mathfrak{X})^2(\Omega). \tag{4.4}$$

Euler's equations of motion are obtained by writing Newton's equations (4.3) in a frame which moves with the body. This special choice of gauge simplifies the equations of motion.

Equation (4.4) expresses a linear relationship between the angular momentum $\mathfrak{L}$ of a particle of mass m at position $\mathfrak{X}$ with angular momentum Ω. Let us call this operator I: $\mathfrak{L} = I(\mathfrak{X})\Omega$. In the case of a rigid body we obtain the total angular momentum by integrating $\mathfrak{L}$ over the body:

$$\mathfrak{L}_{\mathrm{tot}} = \iiint I(\mathfrak{X})\Omega\,d\rho$$

where ρ is the mass density. Since the body is rigid, Ω is constant over the body and

$$\mathfrak{L}_{\mathrm{tot}} = I_{\mathrm{tot}}\Omega$$

where I_{tot} is the linear transformation $\iiint I(\mathfrak{X})\,d\rho$. (Let us now drop the subscript "tot".)

We claim I is a symmetric transformation on $so(3)$ with respect to the inner product $(P, Q) = K(P, Q)$, where K is the Killing form. In fact, at each point $\mathfrak{X}$, $(I(\mathfrak{X})P, Q) = -mK((\mathrm{ad}\,\mathfrak{X})^2(P), Q) = -mK(P, (\mathrm{ad}\,\mathfrak{X})^2(Q))$. Consequently the operator $I = I_{\mathrm{tot}}$ is also symmetric.

Since I is symmetric it is diagonalizeable. Its eigenvalues, denoted by I_j, are called the moments of inertia. Let E_j be a basis for $so(3)$ of eigenvectors of I, so that $IE_j = I_jE_j$. Writing $\Omega = \Omega_jE_j$ we have

$$\mathfrak{L} = I(\Omega) = \sum_{j=1}^{3} I_j\Omega_jE_j.$$

Euler's equations of motion then become

$$\sum_{j=1}^{3} I_j\dot{\Omega}_jE_j + \left[\sum_{k=1}^{3} \Omega_kE_k, \sum_{l=1}^{3} I_l\Omega_lE_l\right] = 0.$$

These reduce to

$$\begin{aligned} I_1\dot{\Omega}_1 &= (I_2 - I_3)\Omega_2\Omega_3 \\ I_2\dot{\Omega}_2 &= (I_3 - I_1)\Omega_3\Omega_1 \\ I_3\dot{\Omega}_3 &= (I_1 - I_2)\Omega_1\Omega_2. \end{aligned} \tag{4.5}$$

We shall return to these equations later. Note that they are really equations of evolution for the matrix $\Omega = O^t\dot{O}$ where $O(t)$ was the transformation to the rotating frame. Since $\Omega(t) \in so(3)$ the Euler equations live in $so(3)$. We shall see later that the corresponding matrix $O(t)$ is a geodesic in $SO(3)$ with an appropriate choice of metric.

1. Let $A \in SU(2)$, $A = \begin{pmatrix} \alpha & \beta \\ -\bar{\beta} & \bar{\alpha} \end{pmatrix}$. Set $\alpha = \chi + i\zeta$ and $\beta = \xi + i\eta$. The parameters ξ, η, ζ, χ are called the Cayley–Klein parameters. The angular momentum matrix is $\Omega = \dot{A}A^* = \omega_1 E_1 + \omega_2 E_2 + \omega_3 E_3$, where $E_j = -\frac{i}{2}\sigma_j$. Compute ω_1, ω_2, ω_3 in terms of the Cayley–Klein parameters.

2. Let $D_t = (d/dt) + \Omega$ and $D'_t = (d/dt) + \Omega'$ be the total derivatives in two different frames O and O' related by $O'B = O$. How are the matrices Ω and Ω' related?

15. The Harmonic Oscillator and Boson Calculus

The harmonic oscillator can be solved by purely algebraic methods. Despite its simplicity it is an important example for quantum mechanics because the electromagnetic field can be treated as a collection of oscillators; moreover, a great many systems such as molecules and solids are well approximated near equilibrium by systems of oscillators.

A harmonic oscillator—a mass on a spring, for example—has one degree of freedom, so is described by a single coordinate q and momentum p. The classical Hamiltonian is

$$H = \frac{1}{2m}(p^2 + m^2\omega^2 q^2)$$

where m is the mass and ω is the frequency. As in Section 13 we quantize this system by replacing p and q by Hermitian operators P' and Q' whose commutation relations are $[Q', P'] = i\hbar\mathbb{1}$. It is convenient to replace P' and Q' by the dimensionless operators $P = P'/\sqrt{mh\omega}$ and $Q = Q'\sqrt{\dfrac{m\omega}{\hbar}}$. Then

$$H = \tfrac{1}{2}\hbar\omega(P^2 + Q^2)$$

and $[Q, P] = i\mathbb{1}$. Planck's constant $\hbar$ has the dimensions (energy) (time) and ω has the dimension (time)$^{-1}$ so $\hbar\omega$ is a unit of energy. Define the operators

$$a = (P - iQ)/\sqrt{2} \quad \text{and} \quad a^* = (P + iQ)/\sqrt{2}$$

as before. Then the Hamiltonian for the quantized oscillator is

$$H = \hbar\omega(a^*a + \tfrac{1}{2})$$

and $[a, a^*] = \mathbb{1}$.

To "solve" the oscillator means to find a representation of the operators a^* and a on a Hilbert space. We shall proceed heuristically and construct a Hilbert space under the (plausible) assumption that H has an eigenvector ψ_λ. (The plausibility of this assumption depends on whether you are a mathematician or a physicist.) We will see later that the Hilbert space we construct is l_2, the space of square summable sequences.

If $H\psi_\lambda = \lambda\hbar\omega\psi_\lambda$ then the commutation relation $[H, a] = -\hbar\omega a$ shows that $a\psi_\lambda$ is also an eigenvector of H with eigenvalue $(\lambda - 1)\hbar\omega$: $H(a\psi_\lambda) = (aH - \hbar\omega a)\psi_\lambda = a(H - \hbar\omega)\psi_\lambda = (\lambda - 1)\hbar\omega(a\psi_\lambda)$. Similarly, $[H, a^*] = \hbar\omega a^*$ and so $a^*\psi_\lambda$ is an eigenvector of H with eigenvalue $(\lambda + 1)\hbar\omega$. Thus the commutation relations imply that a^* and a act as "shift" operators on the eigenvectors of H. Because of these properties a^* is called a "raising" or "creation" operator and a is called a "lowering" or "destruction" operator.

When we apply a to ψ_λ sufficiently often we must eventually get zero, because although a lowers the eigenvalue, these eigenvalues are bounded below. In fact, the inner product of $H\psi$ with ψ is $(H\psi, \psi) = \hbar\omega((a^*a + \frac{1}{2})\psi, \psi) = \hbar\omega(\frac{1}{2}\|\psi\|^2 + \|a\psi\|^2) \geq \frac{1}{2}\hbar\omega\|\psi\|^2$; so the eigenvalues of H are bounded below by $\frac{1}{2}\hbar\omega$. Therefore there must exist an eigenvector ψ_0 with minimum eigenvalue $\lambda_0\hbar\omega$, for which $a\psi_0 = 0$. We call ψ_0 the *ground state* or *vacuum*. Since $a\psi_0 = 0$, $H\psi_0 = \frac{1}{2}\hbar\omega\psi_0$ so $\lambda_0 = \frac{1}{2}$ and the energy of the ground state is $\frac{1}{2}\hbar\omega$.

Let ψ_0 be normalized to unity so that $\|\psi_0\| = 1$. We leave it as an exercise to show that the eigenvectors

$$\psi_n = \frac{(a^*)^n}{\sqrt{n!}}\psi_0$$

are all unit vectors also, and that

$$a^*\psi_n = \sqrt{n + 1}\,\psi_{n+1}, \qquad a\psi_n = \sqrt{n}\,\psi_{n-1}. \tag{4.6}$$

Using this we may construct our representation for the harmonic oscillator as follows. Let $\mathscr{H}$ be the Hilbert space with $\psi_0, \psi_1, \ldots$ as an orthonormal basis. Define the operators a^* and a by (4.6) for $n \geq 1$ with $a\psi_0 = 0$. Then a^* is the adjoint of a, and they satisfy the commutation rule $[a, a^*] = 1$. With $H = \hbar\omega(a^*a + \frac{1}{2})$ we find that $H\psi_n = (n + \frac{1}{2})\hbar\omega\psi_n$ so the ψ_n are indeed the eigenvectors of H. This justifies our heuristic approach, and constitutes a "solution" of the harmonic oscillator problem.

Dirac's notation for states in a Hilbert space is often convenient and widely used in the physics literature. Let $|n\rangle$ denote the normalized eigenvector of a^*a with eigenvalue n. Then

$$H|n\rangle = (n + \frac{1}{2})\hbar\omega|n\rangle$$

$$a|n\rangle = \sqrt{n}|n - 1\rangle \qquad a^*|n\rangle = \sqrt{n + 1}|n + 1\rangle.$$

The vectors in the dual space are denoted by $\langle m|$, and the inner product is $\langle m|n\rangle$. In the case of a complex Hilbert space we require that $\langle \lambda m|n\rangle = \bar{\lambda}\langle m|n\rangle$ but $\langle m|\lambda n\rangle = \lambda\langle m|n\rangle$. We can let a act in either direction. For example

$$\langle m|a^* = (a|m\rangle)^* = \sqrt{m}|m - 1\rangle^* = \sqrt{m}\langle m - 1|;$$

so

$$\langle m|a^*|n\rangle = \sqrt{m}\langle m-1|n\rangle = \sqrt{m}\,\delta_{(m-1)n}$$

if a^* acts to the left, while

$$\langle m|a^*|n\rangle = \sqrt{n+1}\langle m|n+1\rangle = \sqrt{n+1}\,\delta_{m(n+1)}$$

if a^* acts to the right.

The extension of this analysis to a finite collection of independent oscillators is quite simple. We index the oscillators by α and let the frequency of the αth oscillator be ω_α. The Hamiltonian is

$$H = \sum_\alpha \hbar\omega_\alpha(a_\alpha^* a_\alpha + \tfrac{1}{2}).$$

The creation and destruction operators satisfy the commutation relations

$$[a_\alpha, a_\beta] = 0, \qquad [a_\alpha^*, a_\beta^*] = 0, \qquad [a_\alpha, a_\beta^*] = \delta_{\alpha\beta}\mathbb{1}. \tag{4.7}$$

The vacuum state is denoted by $|0\rangle$ and satisfies $a_\alpha|0\rangle = 0$ for all α. The normalized state

$$|n_1, n_2, \ldots\rangle \equiv \frac{(a_1^*)^{n_1}(a_2^*)^{n_2}\ldots|0\rangle}{\sqrt{n_1!\,n_2!\ldots}} \tag{4.8}$$

contains n_1 excitations of the first oscillator, n_2 of the second, and so on. Its energy is $\hbar\omega\sum_\alpha(n_\alpha + \tfrac{1}{2})$. Since $[a_\alpha^*, a_\beta^*] = 0$ the order in which we apply the creation operators is immaterial. Thus, for example, $a_1^* a_2^* a_1^* a_1^*$ acting on $|0\rangle$ is the same as $(a_1^*)^3 a_2^*$ acting on $|0\rangle$; the resulting state is unambiguously labeled $\sqrt{3!}\,|3, 1, 0, 0, \ldots\rangle$. The operators a_α, a_α^* are called *boson operators* and the excitations are called bosons. In the quantized theory of the electromagnetic field the bosons are the photons; in solid state physics the bosons are the quantized vibrations of a lattice and are called phonons; physics abounds with many other bosons as well.

EXERCISES

1. (a) Show that $[a, (a^*)^n] = n(a^*)^{n-1}$.
   (b) Show that $\|(a^*)^n\psi_0\|^2 = n\|(a^*)^{n-1}\psi_0\|^2$.
   (c) Show that $\psi_n = (a^*)^n\psi_0/\sqrt{n!}$ is normalized.

2. Set $(Qf)(x) = xf(x)$ and $(Pf)(x) = (1/i)(\partial/\partial x)f(x)$. P and Q are densely defined, self-adjoint operators on the Hilbert space $L_2(\mathbb{R})$.
   (a) Show that $[Q, P]f = \mathbb{1}$ if f is differentiable.
   (b) Find a^*, and a, and the Hamiltonian.
   (c) Find the ground state ψ_0 such that $a\psi_0 = 0$. What are the higher order states ψ_n?

3. Show that the states (4.8) are normalized to unity provided $|0\rangle$ is.

4. Show that $a_j^* a_j|n_1, n_2, \ldots\rangle = n_j|n_1, n_2 \ldots\rangle$. Thus $a_j^* a_j$ counts the number of bosons in the jth state.

16. Boson Realizations of Lie Algebras

In this section we construct representations of the Lie algebras $su(2)$ and $u(2)$ using the calculus of boson operators. The idea of using creation and annihilation operators to obtain a realization of the Lie algebra $su(2)$ was introduced by Schwinger. Later, in Chapter 12 we construct representations of some semi-simple Lie algebras from this point of view. The advantage of this approach, we feel, is that it lends insight into the structure and representation of semi-simple Lie algebras by linking them to a bona fide physical model. Indeed, thinking in terms of elementary particles enhances ones understanding of the semi-simple Lie algebras.

Let us turn to the construction of representations of $su(2)$ from the calculus of two boson operators a_1, a_2, a_1^*, a_2^* and the harmonic oscillator states $|n_1, n_2\rangle$ given by (4.8). In quantum mechanics, as we have said, the angular momentum operators satisfy the commutation relations of $su(2)$. Let us explain this statement. The components of angular momentum of a classical particle are given by

$$L_j = x^k p_l - x^l p_k \qquad j, k, l \text{ in cyclic order,}$$

where x^j are the components of position and p_j are the components of momentum. These classical observables have the Poisson brackets $\{L_j, L_k\} = \varepsilon_{jkl} L_l$. According to the rules for quantization the corresponding angular momentum operators J_1, J_2, J_3 are to satisfy the commutation relations $[J_j, J_k] = i\hbar\varepsilon_{jkl}J_l$. These operators may be obtained by replacing p_j by $(\hbar/i)(\partial/\partial x_j)$. Then

$$J_j = \frac{\hbar}{i}\left(x^k\frac{\partial}{\partial x^l} - x^l\frac{\partial}{\partial x^k}\right)$$

where j, k, l are in cyclic order.

Let us now drop the $\hbar$ (for example, by choosing a system of units in which $\hbar = 1$) and take as the commutation relations for $su(2)$ $[J_j, J_k] = i\varepsilon_{jkl}J_l$. Define $J_\pm = (J_1 \pm iJ_2)$ and $J_0 = J_3$. Then J_+, J_-, J_0 satisfy the commutation relations

$$[J_0, J_\pm] = \pm J_\pm \qquad [J_+, J_-] = 2J_0. \qquad (4.9)$$

Let a_1, a_2, a_1^* and a_2^* be boson operators satisfying the commutation relations (4.7) and put

$$J_+ = a_1^* a_2, \qquad J_- = a_2^* a_1, \qquad J_0 = \tfrac{1}{2}(a_1^* a_1 - a_2^* a_2).$$

It is easily checked that the operators J_+, J_- and J_0 defined in this way satisfy the commutation relations (4.9), and thus constitute a boson realization of $su(2)$.

We may further use these boson operators to construct representations of $su(2)$. Consider the action of J_+, J_- and J_0 on the harmonic oscillator states $|n_1, n_2\rangle$ of (4.8). We find

$$J_+|n_1, n_2\rangle = a_1^* a_2 |n_1, n_2\rangle$$
$$= \sqrt{(n_1 + 1)n_2}\, |n_1 + 1, n_2 - 1\rangle$$
$$J_-|n_1, n_2\rangle = \sqrt{n_1(n_2 + 1)}\, |n_1 - 1, n_2 + 1\rangle$$
$$J_0|n_1, n_2\rangle = \tfrac{1}{2}(n_1 - n_2)|n_1, n_2\rangle.$$

Then J_+ creates an extra particle in the first state while destroying one in the second state, while J_- reverses this process. J_0 leaves the number in each state alone. Note that the total number of particles is conserved by each of J_+, J_- and J_0. That is, these operators leave invariant the subspaces $\mathscr{P}_N = \{|n_1, n_2\rangle | n_1 + n_2 = N\}$. The dimension of $\mathscr{P}_N$ is $N + 1$, so we have an $N + 1$ dimensional representation of $su(2)$.

An "obvious" operator to consider is $P = a_1^* a_1 + a_2^* a_2$. It simply counts the total number of particles: $P|n_1, n_2\rangle = (n_1 + n_2)|n_1, n_2\rangle$. Moreover, P commutes with J_+, J_- and J_0.

In applications to angular momentum problems it is more convenient to label the states in a different way. Let $m = \tfrac{1}{2}(n_1 - n_2)$ and $j = \tfrac{1}{2}(n_1 + n_2)$. Thus we label the state $|n_1, n_2\rangle$ by $|j, m\rangle = \left|\dfrac{n_1 + n_2}{2}, \dfrac{n_1 - n_2}{2}\right\rangle$, and

$$|j, m\rangle = \frac{(a_1^*)^{j+m}(a_2^*)^{j-m}}{\sqrt{(j + m)!(j - m)!}}|0\rangle. \tag{4.10}$$

In this labelling, the action of J_+, J_-, J_0 is given by

$$J_+|j, m\rangle = \sqrt{(j + m + 1)(j - m)}\,|j, m + 1\rangle$$
$$J_-|j, m\rangle = \sqrt{(j + m)(j - m + 1)}\,|j, m - 1\rangle \tag{4.11}$$
$$J_0|j, m\rangle = m|j, m\rangle$$
$$P|j, m\rangle = 2j|j, m\rangle.$$

Thus P and J_0 are diagonal matrices in this basis while J_+ and J_- are shift operators. The dimension of this representation is $2j + 1$, m is called the magnetic quantum number, and the set of $(2j + 1)$ states $|j, m\rangle$, $m = -j$, $-j + 1, \ldots j$ is called a $2j + 1$ *multiplet*. Note that j is either an integer or a half-integer.

There are other applications of the $su(2)$ model in physics than to angular momentum. In nuclear physics the neutron and proton are treated as two states of an $su(2)$ doublet, the nucleon. (Heisenberg, 1932). In this case j is called the isospin and labeled I, while m is labeled I_3 or I_z and called the z-component of isospin, in analogy with the angular momentum theory.

One final remark must be made regarding the representations of $su(2)$ which were constructed above. In the construction of the states $(a_1^*)^{n_1}(a_2^*)^{n_2}|0\rangle$ the order in which the boson operators operate is immaterial. In labeling the states it is only necessary to count the total number of bosons of type 1 and type 2. This will not be the case when we wish to construct representations of other Lie algebras.

The evolution of particle physics has led to the introduction of larger and more complicated Lie groups, such as $SU(4)$ in nuclear physics, $SU(3)$ for the strong interactions, and $SU(2) \times U(1)$ for the weak and electromagnetic interaction. Some of these developments will be discussed in Chapter 12. For now we conclude this chapter with a discussion of the Lie algebra $u(2) \cong su(2) \oplus u(1)$, for it is not only pertinent in physics but also illustrates further aspects of representation theory.

A realization of $u(2)$ is obtained by adding P to J_+, J_- and J_0 in the representation of $su(2)$ above. These operators indeed form $u(2)$ but the representations of $u(2)$ given by the states $\{|n_1, n_2\rangle\}$ are precisely the $su(2)$ representations. They have a "new" label, the eigenvalue of P, but this is just $2j = n_1 + n_2 = N$. New representations are obtained, however, if we introduce a second set of independent bosons b_1, b_2, b_1^* and b_2^* which commute with the a_α and a_α^*. The vacuum $|0\rangle$ is annihilated by b_1 and b_2 as well as a_1 and a_2. Let

$$P = a_1^* a_1 + a_2^* a_2 + b_1^* b_1 + b_2^* b_2$$

$$J_+ = a_1^* a_2 + b_1^* b_2$$

$$J_- = a_2^* a_1 + b_2^* b_1$$

$$J_0 = \tfrac{1}{2}(a_1^* a_1 - a_2^* a_2) + \tfrac{1}{2}(b_1^* b_1 - b_2^* b_2).$$

It is immediately seen that these operators satisfy the same commutation rules as before and so they provide another realization of $u(2)$. But now we can construct new states in which the eigenvalues of P and those of J_0 are independent. For example, the state $(a_1^* b_2^* - a_2^* b_1^*)^\Lambda |0\rangle$ is annihilated by J_+, J_- and J_0 so it is an $su(2)$ singlet. That is, it gives a one-dimensional (trivial) representation of $su(2)$. But it is an eigenvector of P with eigenvalue 2Λ. By contrast, the only singlet of $u(2)$ which we obtained without the b's was the vacuum state $|0\rangle$. Now we have an infinite set of distinct one dimensional, non-trivial representations of $u(2)$.

More generally, the states

$$|\Lambda, n_1, n_2\rangle = N_{\Lambda n_1 n_2}(a_1^* b_2^* - a_2^* b_1^*)^\Lambda (a_1^*)^{n_1}(a_2^*)^{n_2}|0\rangle$$

where the normalization constant is given by

$$N_{\Lambda n_1 n_2}^2 = \frac{(n_1 + n_2)!}{n_1! n_2!} \frac{n_1 + n_2 + 1}{\Lambda!(\Lambda + n_1 + n_2 + 1)!}$$

provide representations of $u(2)$ labeled by the eigenvalues of P and the maximum eigenvalue of J_0. The eigenvalue of J_0 is $m = \tfrac{1}{2}(n_1 - n_2)$ as before so $|\Lambda, j, m\rangle$ is an alternative labeling. For each value of Λ we get a different $2j + 1$ dimensional representation of $u(2)$.

The construction of additional representations of $u(2)$ by this device illustrates an important point in the theory of representations. The operator

$a_1^* b_2^* - a_2^* b_1^*$ is anti-symmetric under the interchange $1 \leftrightarrow 2$. Such an anti-symmetric expression could never be formed with a single set of boson operators: $a_1^* a_2^* - a_2^* a_1^* = 0$ since a_1^* and a_2^* commute. In order to construct all representations of a general semi-simple Lie algebra several bosons a_i, b_i, c_i must be introduced, and the representations of the symmetric group play a fundamental (albeit sometimes subtle) role. In order to obtain all the representations of $u(2)$ we had to consider anti-symmetric states.

EXERCISES

1. Show that $P|\Lambda \, n_1 \, n_2\rangle = (2\Lambda + n_1 + n_2)|\Lambda \, n_1 \, n_2\rangle$.

2. Show that $[J, (a_1^* b_2^* - a_2^* b_1^*)] = 0$ for $J = J_+, J_-$ or J_0 and hence that

$$J_+|\Lambda n_1 n_2\rangle = \sqrt{(n_1 + 1)(n_2)}|\Lambda \, n_1 + 1 \, n_2 - 1\rangle$$
$$J_-|\Lambda n_1 n_2\rangle = \sqrt{n_1(n_2 + 1)}|\Lambda \, n_1 - 1 \, n_2 + 1\rangle$$
$$J_0|\Lambda n_1 n_2\rangle = \tfrac{1}{2}(n_1 - n_2)|\Lambda \, n_1 \, n_2\rangle.$$

   Then the representation of u(2) with given Λ, $N = n_1 + n_2$, is exactly the $N + 1$ dimensional of $su(2)$ we found earlier.

3. Show that $K_+ = a_1^* a_2^*$, $K_- = a_1 a_2$, $K_0 = \tfrac{1}{2}(a_1^* a_1 + a_2 a_2^*)$ generate the Lie algebra (2.3). Find the action of these operators on the states $|n_1, n_2\rangle$ and show that these are infinite dimensional representations.

4. Construct the Lie algebra $su(3)$ using a_1, a_2, a_3 and their conjugates.

DIFFERENTIAL GEOMETRY AND LIE GROUPS

CHAPTER 5

Calculus on Manifolds

17. Vector Fields, Flows, and 1-Forms

The equations of mathematical physics are typically ordinary or partial differential equations for vector or tensor fields over Riemannian manifolds whose group of isometries is a Lie group. It is taken as axiomatic that the equations be independent of the observer, in a sense we shall make precise below; and the consequence of this axiom is that the equations are invariant with respect to the group action. The action of a Lie group on tensor fields over a manifold is thus of primary importance. The action of a Lie group on a manifold M induces in a natural way automorphisms of the algebra of C^∞ functions over M and on the algebra of tensor fields over M. The one parameter subgroups of the group induce one parameter subgroups of automorphisms of the tensor fields. The infinitesimal generators of these groups of automorphisms are the Lie derivatives of the action.

For example the Navier–Stokes equations for a viscous incompressible fluid are

$$\Delta u^i - \frac{\partial p}{\partial x^i} = u^j \frac{\partial u^i}{\partial x^j} + \frac{\partial u^i}{\partial t}$$

$$\frac{\partial u^i}{\partial x^i} = 0.$$

(We use the summation convention here.) This is a system of equations for a vector field $u = (u^i)$ and scalar p. One may ask how these equations transform under rigid motions of the underlying space $\mathbb{R}^3$. In fact, one finds that these equations are equivariant with respect to the group of rigid motions $\mathscr{E}(3)$ of $\mathbb{R}^3$ in the following sense. A rigid motion in $\mathbb{R}^3$ is given by $\{O, a\} X = OX + a$

where $O \in 0(3)$ and $a \in \mathbb{R}^3$. A representation of $\mathscr{E}(3)$ on the four component quantity $w = (u(x), p(x))$ is given by

$$
T_g \begin{pmatrix} u^1 \\ u^2 \\ u^3 \\ p \end{pmatrix}(x) = \begin{pmatrix} 0_{11} & 0_{12} & 0_{13} & 0 \\ 0_{21} & 0_{22} & 0_{23} & 0 \\ 0_{31} & 0_{32} & 0_{33} & 0 \\ 0 & 0 & 0 & 1 \end{pmatrix} \begin{pmatrix} u^1 \\ u^2 \\ u^3 \\ p \end{pmatrix} (g^{-1}x)
$$

where $g = \{0, a\}$. Now let $G = (G^1, G^2, G^3, G^4)$ be the four components of the Navier–Stokes equations; that is, $G^i = \Delta u^i - \dfrac{\partial p}{\partial x^i} - u^j \dfrac{\partial u^i}{\partial x^j}$ for $i = 1, 2, 3$, and $G^4 = \displaystyle\sum_{i=1}^{3} \dfrac{\partial u^i}{\partial x^i}$. The Navier–Stokes equations are equivariant with respect to the group of rigid motions in the sense that

$$
T_g G(w) = G(T_g w).
$$

This mathematical condition is a statement of the fact that the Navier–Stokes equations are the same in any Euclidean frame of reference.

As a second example, the electron has an internal structure due to its spin, and so is represented by a vector valued wave function known as a spinor. In the non-relativistic theory the spinor has two components. We saw in Chapter 1 that a matrix $A \in SU(2)$ generates a rotation $R(A)$ in $SO(3)$. Under the action of a rotation R, $x' = Rx$, and the spinor ψ transforms according to the rule

$$
\psi'(x') = e^{-i(\theta/2)\hat{n}\cdot\vec{\sigma}} \psi(x)
$$

where R is the rotation through an angle θ about the axis $\hat{n}$. Note that $(T_A\psi)(x) = A\psi(R^{-1}(A)x)$ is a representation of $SU(2)$. In Dirac's relativistic theory of the election the spinor has 4 components and transforms according to a representation of the Lorentz group. The Dirac equations of the electron are equivariant with respect to that representation. Alternatively, we may say the Dirac equations are invariant under Lorentz transformations.

In §17 and §18 we present a quick introduction to the basic ideas of tensor calculus on manifolds, turning, in §19, to the transformations of tensor fields induced by group actions on the manifold. We begin with an introduction to vector fields and flows in $\mathbb{R}^n$ and then discuss the situation on a general differentiable manifold.

Let us begin with a simple example. Consider the transformations of the plane given by

$$
\varphi_t(x, y) = (x \cos t + y \sin t, -x \sin t + y \cos t).
$$

The family $\{\varphi_t\}$ constitutes a one parameter group of transformations of the plane, since, as we easily see, $\varphi_{t+s} = \varphi_t \circ \varphi_s$. Such a one-parameter transformation group is called a *flow* on $\mathbb{R}^2$. We can easily find a set of differential equations for this flow. Setting

$$x(t) = \varphi_t^1(x, y) = x \cos t + y \sin t$$

$$y(t) = \varphi_t^2(x, y) = -x \sin t + y \cos t$$

we find that x and y satisfy the differential equations

$$\frac{dx}{dt} = y \qquad \frac{dy}{dt} = -x.$$

These ordinary differential equations are said to generate the flow.

Now let $C^\infty(\mathbb{R}^2)$ denote the algebra of infinitely differentiable functions on $\mathbb{R}^2$. The flow φ_t induces a one-parameter group of automorphisms ρ_t of $C^\infty(\mathbb{R}^2)$ defined by $\rho_t f = f \circ \varphi_t$. It is immediate that

(i) ρ_t is linear on $C^\infty(\mathbb{R}^2)$
(ii) $\rho_t(fg) = (\rho_t f)(\rho_t g)$
(iii) $\rho_{t+s} = \rho_t \rho_s$.

As we showed in §12, the infinitesimal generator of a one-parameter group of automorphisms of an algebra is a derivation on the algebra. We calculate the infinitesimal generator of ρ_t as follows. Putting $f_t = \rho_t f$ we have

$$\frac{\partial f_t}{\partial t} = \frac{d}{dt} f(x(t), y(t))$$

$$= \frac{\partial f}{\partial x}\frac{dx}{dt} + \frac{\partial f}{\partial y}\frac{dy}{dt}$$

$$= y\frac{\partial f}{\partial x} - x\frac{\partial f}{\partial y},$$

and thus $R = y(\partial/\partial x) - x(\partial/\partial y)$ is the infinitesimal generator of ρ_t. We note that R is indeed a derivation of $C^\infty(\mathbb{R}^2)$: it is linear on $C^\infty(\mathbb{R}^2)$ and $R(fg) = (Rf)g + f(Rg)$.

Conversely, given R we can find the differential equations of the flow. We simply take for our function f the component functions x and y (which are certainly elements of $C^\infty(\mathbb{R}^2)$) and solve $\partial f_t/\partial t = Rf$. We get

$$\frac{d}{dt}x(t) = Rx = y, \qquad \frac{d}{dt}y(t) = Ry = -x.$$

It is clear that the above discussion can be carried over quite generally to $\mathbb{R}^n$. A flow on $\mathbb{R}^n$ is a one-parameter transformation group φ_t such that $\varphi_{t+s} = \varphi_t \circ \varphi_s$ and $\varphi_0 = 1$. (One could also define a local flow on a domain $U \subset \mathbb{R}^n$ and a subinterval $|t| \leq \delta$; all statements about the composition property would have to be modified accordingly.) Any such flow can be obtained by integrating a system of ordinary differential equations, as follows. Define the components

$$X^i(x) = \frac{d}{dt}\varphi_t^i(x)|_{t=0}.$$

Setting $x^i(t) = \varphi_t^i(x)$, we get

$$\frac{dx^i}{dt} = \lim_{s \to 0} \frac{\varphi_{t+s}^i - \varphi_t^i}{s}$$

$$= \lim_{s \to 0} \frac{\varphi_s^i - 1}{s} \circ \varphi_t$$

$$= X^i \circ \varphi_t(x)$$

$$= X^i(x(t)).$$

Conversely, any smooth set of functions X^i generates a flow by integrating the above system of differential equations.

As before, the flow φ_t induces a one-parameter group of automorphisms ρ_t on $C^\infty(\mathbb{R}^n)$ given by $\rho_t f = f \circ \varphi_t$. Setting $f_t = \rho_t f$ we find

$$\frac{d}{dt} f_t = \frac{d}{dt} f(x^i(t)) = \frac{\partial f}{\partial x^i} \frac{dx^i}{dt} = X^i \frac{\partial f}{\partial x^i}.$$

Thus the infinitesimal generator of ρ_t is the derivation

$$X = X^i \frac{\partial}{\partial x^i}.$$

We identify such a first order partial differential operator with a vector field. We may think of X as a field of directional derivatives. Treating a vector field as a first order operator rather than as a set of components $(X^1 \ldots X^n)$ has its advantages. For one thing, such a representation automatically keeps track of the transformation laws of the components X^i under coordinate transformations (see §19).

The above arguments can also be carried over to a smooth manifold M by simply assuming that $x^1 \ldots x^n$ are a system of local coordinates on M. A flow φ_t on M induces a one parameter group of automorphisms ρ_t of $C^\infty(M)$ whose infinitesimal generator X, in local coordinates, is given by $Xf = X^i(\partial f/\partial x^i)$.

It is obvious that every operator $X = X^i(\partial/\partial x^i)$ is a derivation on $C^\infty(M)$, where $x^1 \ldots x^n$ are local coordinates on M. Conversely

Theorem 5.1. Let X be a derivation of $C^\infty(M)$ and let $x^1 \ldots x^n$ be a system of local coordinates on M. Then $Xf = \sum_{j=1}^{n} X^j(x) \frac{\partial f}{\partial x^j}$, where $X^j(x) = Xx^j$.

Remark. The coordinate functions $x^1 \ldots x^n$ are themselves functions on M, so Xx^j is defined as the derivation X acting on the coordinate functions x^j.

PROOF. Since X is a derivation, $X(1) = X(1 \cdot 1) = 1 \cdot X(1) + X(1) \cdot 1 = 2X(1)$; hence $X(1) = 0$ and X annihilates all constant functions. Let the coordinates $x^1 \ldots x^n$ be valid in a domain U of M and let f be any smooth function on U. By Taylor's theorem

$$f(x) = f(x_0) + \sum_{j=1}^{n} \frac{\partial f}{\partial x^j}(x_0)(x^j - x_0^j) + R_j(x)(x^j - x_0^j)$$

where $R_j(x)$ is a set of smooth functions which vanish at $x = x_0$. Now $X(x^j - x_0^j)R_j(x) = (X(x^j - x_0^j))R_j + (x^j - x_0^j)XR_j = 0$ since both $x^j - x_0^j$ and $R^j(x)$ vanish at x_0. Therefore

$$Xf = Xf(x_0) + \sum_{j=1}^{n} X(x^j - x_0^j)\frac{\partial f}{\partial x^j}(x_0)$$

$$= \sum_{j=1}^{n} (X^j(x_0))\frac{\partial f}{\partial x^j}(x_0)$$

where $X^j(x_0) = (Xx^j)|_{x=x_0}$. $\square$

Theorem 5.1 shows that vector fields and derivations are one and the same thing, or, more precisely, that vector fields are derivations computed in local coordinates. From the theorem it follows that if $x^1 \dots x^n$ are local coordinates on a domain $U \subset M$ then $\partial/\partial x^1, \dots, \partial/\partial x^n$ form a basis for the vector fields on U. We may identify the vector fields acting at a point $p \in M$ with the tangent vectors to M at p (i.e. with tangential differentiations to M at p). The tangent space of an n-dimensional manifold is thus n-dimensional.

A vector field on a manifold M is thus regarded as a derivation of $C^\infty(M)$, and vice-versa. The derivations of M are denoted by $\mathcal{D}^1(M)$. Thus $X \in \mathcal{D}^1(M)$ if X is a linear map from $C^\infty(M)$ to $C^\infty(M)$ and $X(fg) = (Xf)g + f(Xg)$. Such a definition is intrinsic (can be expressed independently of local coordinate systems) and furthermore does not require the manifold M to be embedded in some ambient space. For example, by this definition we can talk about the Lie algebra of a Lie group $\mathfrak{G}$ as being the tangent space at the identity without *a priori* requiring $\mathfrak{G}$ to be realized as a matrix group.

From Theorem 5.1 we have

Theorem 5.2. *Every derivation X on M induces a flow on M; and conversely, every flow on M is generated by a derivation.*

PROOF. If φ_t is a flow on M then $\rho_t f = f \circ \varphi_t$ is a one-parameter group of automorphisms of $C^\infty(M)$. By the standard argument of §12, $Xf = (d/dt)f \circ \varphi_t|_{t=0}$ is a derivation on $C^\infty(M)$. Conversely, if X is derivation on $C^\infty(M)$ we obtain a flow by introducing local coordinates $x^1 \dots x^n$ on M and solving the ordinary differential equations

$$\frac{dx^i}{dt} = X^i(x).$$

The solution expressed by $x(t) = \varphi_t(x)$ is then a flow on M. $\square$

We now turn our discussion to one-forms. The integral

$$\int \vec{F} \cdot d\vec{x} = \int F_i \, dx^i$$

is called a *line integral*, and the integrand $\omega = F_i \, dx^i$ is called a *one-form*. Just as a vector field was defined as a derivation, a one-form may be defined, in a coordinate free manner, as a linear functional on the derivations. In other words, one-forms and vector fields are in duality. If ω is a one-form we require

$$\omega(fX + gY) = f\omega(X) + g\omega(Y),$$

for X, $Y \in \mathscr{D}^1(M)$ and f and g in $C^\infty(M)$. We denote the one-forms on a manifold M by $\Lambda_1(M)$.

One way to construct a one-form ω is by a duality argument: for $f \in C^\infty$ we define df to be the one-form given by $df(X) = Xf$. Since the local coordinate functions $x^1 \ldots x^n$ are C^∞ functions on M, we can define $dx^i(X) = Xx^i$. In particular, $dx^i(\partial/\partial x^j) = \partial/\partial x^j(x^i) = \delta^i_j$, so the one-forms $dx^1 \ldots dx^n$ form a dual basis to the vector fields $\partial/\partial x^1 \ldots \partial/\partial x^n$. For a general one-form ω we have, in local coordinates,

$$\omega(X) = \omega\left(X^j \frac{\partial}{\partial x^j}\right) = X^j \omega_j$$

where the functions $\omega_j = \omega(\partial/\partial x^j)$ are the components of ω in the coordinate system $x^1 \ldots x^n$. We may therefore write any one-form as $\omega = \omega_j \, dx^j$.

The integral of a one form ω over a path C is defined as follows. Let $\gamma(t)$, $0 \le t \le 1$ be a parametrization of C, and let $\dot{\gamma}$ be the tangent vector to C. Then

$$\int_C \omega = \int_0^1 \omega(\dot{\gamma}) \, dt.$$

In local coordinates $x^1 \ldots x^n$ we have $\omega = \omega_i \, dx^i$ and $\dot{\gamma} = (dx^i/dt)(\partial/\partial x^i)$, so

$$\omega(\dot{\gamma}) = \omega_i \, dx^i \left(\frac{dx^j}{dt} \frac{\partial}{\partial x^j}\right) = \omega_j \frac{dx^j}{dt},$$

and

$$\int_C \omega = \int_0^1 \omega_j(x(t)) \frac{dx^j}{dt} \, dt.$$

The integral on the right is, of course, independent of the parametrization.

EXERCISES

1. The tangent bundle of $\mathbb{R}^3$ can be represented by pairs (X, V) where $X \in \mathbb{R}^3$ and $V \in \mathbb{R}^3$ is a tangent vector at X. Under a rigid motion $g = \{O, a\}$ this pair is transformed to $g(X, V) = (OX + a, OV)$. Show this action is given by the matrix representation

$$g = \begin{pmatrix} 1 & 0 & 0 & 0 \\ a_1 & 0_{11} & 0_{12} & 0_{13} \\ a_2 & 0_{21} & 0_{22} & 0_{23} \\ a_3 & 0_{31} & 0_{32} & 0_{33} \end{pmatrix}, \quad X = \begin{pmatrix} 1 \\ x_1 \\ x_2 \\ x_3 \end{pmatrix}, \quad V = \begin{pmatrix} 0 \\ v_1 \\ v_2 \\ v_3 \end{pmatrix},$$

$$(X, V) = \begin{pmatrix} 1 & 0 \\ x_1 & v_1 \\ x_2 & v_2 \\ x_3 & v_3 \end{pmatrix}.$$

2. The *frame bundle* of $\mathbb{R}^3$ is the set of points X and the set of all orthonormal bases E_1, E_2, E_3 based at X: $E_i \cdot E_j = \delta_{ij}$. Show that the frame bundle of $\mathbb{R}^3$ can be identified with the group of rigid motions of $\mathbb{R}^3$, and that the action of the rigid motions on the frame bundle is the left action of the group on itself.

3. What is the frame bundle of the unit sphere in $\mathbb{R}^3$? Show how to identify the frame bundle with the group of isometries.

18. Differential Forms and Integration

Just as a one-form is the natural integrand for a line integral, a p-form is the natural integrand for a p-dimensional integral. Integration theory in higher dimensions is simplified considerably by the exterior calculus—the calculus of differential forms. Furthermore, many quantities in physics are naturally expressed as differential forms. For example, the electromagnetic field tensor

$$F_{\mu\nu} = \frac{\partial A_\nu}{\partial x^\mu} - \frac{\partial A_\mu}{\partial x^\nu} = \begin{pmatrix} 0 & -E_1 & -E_2 & -E_3 \\ E_1 & 0 & B_3 & -B_2 \\ E_2 & -B_3 & 0 & B_1 \\ E_3 & B_2 & -B_1 & 0 \end{pmatrix}$$

is a skew symmetric ($F_{\mu\nu} = -F_{\nu\mu}$) second order covariant tensor. In the language of differential forms we say $F_{\mu\nu}$ are the components of a two form and write $F = F_{\mu\nu} dx^\mu \wedge dx^\nu$.

The symbol $\wedge$ denotes the wedge or exterior product, which is an anti-symmetric product. For example, the wedge product of two one-forms ω_1 and ω_2 is the two-form

$$(\omega_1 \wedge \omega_2)(X, Y) = \tfrac{1}{2}(\omega_1(X)\omega_2(Y) - \omega_1(Y)\omega_2(X)).$$

We see that $\omega_1 \wedge \omega_2$ is a skew-symmetric bilinear functional of pairs of vectors X and Y, *viz.*

$$(\omega_1 \wedge \omega_2)(\alpha X + \beta Y, Z) = \alpha(\omega_1 \wedge \omega_2)(X, Z) + \beta(\omega_1 \wedge \omega_2)(Y, Z),$$

$$(\omega_1 \wedge \omega_2)(X, \alpha Y + \beta Z) = \alpha(\omega_1 \wedge \omega_2)(X, Y) + \beta(\omega_1 \wedge \omega_2)(X, Z),$$

$$(\omega_1 \wedge \omega_2)(X, Y) = -(\omega_1 \wedge \omega_2)(Y, X)$$

where $\alpha, \beta \in C^\infty(M)$. It is clear that $\omega_1 \wedge \omega_2 = -\omega_2 \wedge \omega_1$ and that $(\alpha\omega_1 + \beta\omega_2) \wedge \omega_3 = \alpha\omega_1 \wedge \omega_3 + \beta\omega_2 \wedge \omega_3$. Since $\omega_1 \wedge \omega_2 = -\omega_2 \wedge \omega_1$ it follows that $\omega \wedge \omega = 0$ for any one-form ω.

In Cartesian coordinates in the x-y plane, for example, $dx \wedge dy$ should be regarded as the (oriented) area element. In polar coordinates, $x = r\cos\theta$, $y = r\sin\theta$, and we have

$$dx = -r\sin\theta \, d\theta + \cos\theta \, dr$$

$$dy = r\cos\theta \, d\theta + \sin\theta \, dr$$

and

$$dx \wedge dy = (-r\sin\theta \, d\theta + \cos\theta \, dr) \wedge (r\cos\theta \, d\theta + \sin\theta \, dr)$$

$$= r\cos^2\theta \, dr \wedge d\theta - r\sin^2\theta \, d\theta \wedge dr$$

$$= r \, dr \wedge d\theta$$

since $dr \wedge dr = 0$, $d\theta \wedge d\theta = 0$, and $d\theta \wedge dr = -dr \wedge d\theta$.

More generally, a p-form is a multilinear skew symmetric functional of p vector fields. If ω is a p-form and $X_1 \ldots X_p$ are smooth vector fields, then $\omega(X_1 \ldots X_p)$ is a smooth function. Moreover, ω is linear in each variable over the ring of functions. This means that for every j

$$\omega(\ldots fX_j + gX'_j, \ldots) = f\omega(\ldots X_j \ldots) + g\omega(\ldots X'_j \ldots)$$

where f and g belong to $C^\infty(M)$. Moreover, $\omega(X_1 \ldots X_p)$ changes sign when any two entries are interchanged, just like a determinant.

The set of all p-forms on a manifold M is denoted by $\Lambda_p(M)$; functions on M are regarded as 0-forms, so that $\Lambda_0(M) = C^\infty(M)$. The collection of all forms on M is denoted by $\Lambda(M)$. The wedge product of p 1-forms $\omega_1 \ldots \omega_p$ is given by

$$(\omega_1 \wedge \cdots \wedge \omega_p)(X_1 \ldots X_p) = \frac{1}{p!} \det \|\omega_i(X_j)\|.$$

A basis for the p-forms on M in local coordinates is the set $\{dx^{i_1} \wedge \cdots \wedge dx^{i_p} | 1 \leq i_1 < i_2, \ldots < i_p \leq n\}$, so the dimension of Λ_p is $\binom{n}{p}$. There can be no p-forms for $p > \dim M$ by the antisymmetry of the wedge product.

A basis for the two forms on $\mathbb{R}^3$, for example, is given by $dx \wedge dy, dx \wedge dz, dy \wedge dz$. The general two form is given by $A \, dx \wedge dy + B \, dy \wedge dz + C \, dz \wedge dx$ where $A, B, C \in C^\infty(\mathbb{R}^3)$. The most general three form is $f \, dx \wedge dy \wedge dz$, with $f \in C^\infty(\mathbb{R}^3)$.

The wedge product is extended to all forms in $\Lambda(M)$ by

associativity: $(\omega \wedge \mu) \wedge \nu = \omega \wedge (\mu \wedge \nu)$

and

linearity: $(f\omega + g\mu) \wedge v = f\omega \wedge v + g\mu \wedge v,$

$\omega \wedge (f\mu + gv) = f\omega \wedge \mu + g\omega \wedge v.$

An immediate consequence of the associativitity is that $\omega \wedge \mu = (-1)^{pq}\mu \wedge \omega$ for $\omega \in \Lambda_p$ and $\mu \in \Lambda_q$. Then $\Lambda(M)$ is an associative ring with the exterior product as multiplication.

Now let us define the exterior derivative d of a differential form. We have already seen how to define d on $\Lambda_0(M) = C^\infty(M)$: namely $(df)(X) = Xf$. The operator d is extended to all p-forms by the following properties

(i) $d^2 = 0$
(ii) $d(\omega \wedge \mu) = d\omega \wedge \mu + (-1)^p \omega \wedge d\mu$ for $\omega \in \Lambda_p$
(iii) d is a linear map from Λ_p to Λ_{p+1}.

These properties uniquely determine d. (See, for example, Singer and Thorpe.)
For example, in $\mathbb{R}^n$,

$$d(a_j(x)\, dx^j) = (da_j) \wedge dx^j + a_j \wedge d(dx^j)$$

$$= da_j \wedge dx^j$$

$$= \frac{\partial a_j}{\partial x^k} dx^k \wedge dx^j$$

$$= \sum_{k<j} \left(\frac{\partial a_j}{\partial x^k} - \frac{\partial a_k}{\partial x^j} \right) dx^k \wedge dx^j.$$

In $\mathbb{R}^3$

$$d(A\, dy \wedge dz + B\, dz \wedge dx + C\, dx \wedge dy) = \left(\frac{\partial A}{\partial x} + \frac{\partial B}{\partial y} + \frac{\partial C}{\partial z} \right) dx \wedge dy \wedge dz.$$

The reader should note the resemblence of these operations to the curl and divergence of vector analysis. Thus the action of d on a 1-form resembles a curl, and d acting on a two-form in $\mathbb{R}^3$ resembles the divergence of a vector field. Now div and curl are operations on vector fields while the exterior derivative d acts on forms. As we shall explain in §19, the operations of vector analysis are tied to the Euclidean metric of $\mathbb{R}^3$, whereas the exterior derivative is not.

The integral of a p-form over a p-dimensional submanifold S is computed by parametrizing S, just as in the case of the line integral. For example, let S be a two dimensional submanifold parametrized by

$$x^i = x^i(u^1, u^2), \qquad i = 1, \ldots, n$$

$$(u^1, u^2) \subset U$$

and let $\omega = \omega_{ij}\, dx^i \wedge dx^j$. Then for each pair i, j,

$$dx^i \wedge dx^j = \frac{\partial x^i}{\partial u^m} du^m \wedge \frac{\partial x^j}{\partial u^n} du^n$$

$$= \frac{\partial x^i}{\partial u^m} \frac{\partial x^j}{\partial u^n} du^m \wedge du^n$$

$$= \frac{\partial(x^i, x^j)}{\partial(u^1, u^2)} du^1 \wedge du^2$$

since m and n run over the integers 1, 2. The expression $\partial(x^i, x^j)/\partial(u^1, u^2)$ is the Jacobian of the mapping $(u^1, u^2) \to (x^i(u^1, u^2), x^j(u^1, u^2))$. We thus have

$$\int_S \omega = \int_U \omega_{ij}(x(u)) \frac{\partial(x^i, x^j)}{\partial(u^1, u^2)} du^1 \wedge du^2.$$

The integral on the right is an ordinary integral over a two dimensional region U in the plane. We must check that this definition of the integral is independent of the parametrization. Under a change of variables $u^j = u^j(v^1, v^2)$ we have

$$\frac{\partial(x^i, x^j)}{\partial(u^1, u^2)} du^1 \wedge du^2 = \frac{\partial(x^i, x^j)}{\partial(u^1, u^2)} \frac{\partial(u^1, u^2)}{\partial(v^1, v^2)} dv^1 \wedge dv^2$$

$$= \frac{\partial(x^i, x^j)}{\partial(v^1, v^2)} dv^1 \wedge dv^2$$

by the chain rule. The independence follows from the usual change of variable formula for multiple integrals. The integral of a p-form over a p-dimensional submanifold is defined in an analogous way.

The familiar theorems of Stokes, Green and Gauss of vector analysis take a unified form in the language of differential forms, *viz.*

Theorem 5.3 (Stokes' Theorem).

$$\int_S d\omega = \int_{\partial S} \omega \tag{5.1}$$

where S is an oriented p-dimensional submanifold and ∂S is its (compatibly) oriented boundary.

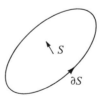

For example, if S is a region in the plane with boundary ∂S then

$$\int_{\partial S} P\,dx + Q\,dy = \int\int_S d(P\,dx + Q\,dy)$$

$$= \int\int_S (Q_x - P_y)\,dx \wedge dy,$$

which is the familiar Green's theorem. The Gauss divergence theorem is obtained by taking $\omega = A\,dy \wedge dz + B\,dz \wedge dx + C\,dx \wedge dy$. Then $d\omega = (A_x + B_y + C_z)\,dx \wedge dy \wedge dz$. We leave it to the reader to show that the integral of ω over a surface S is given by

$$\int_S \omega = \int_S \vec{F} \cdot \hat{n}\,ds$$

where n is the outward unit normal, and $\vec{F} = (A, B, C)$ (see exercise 2).

EXERCISES

1. Let $x^1 \ldots x^n$ be local coordinates in M and let S be a submanifold parametrized by $u^1 \ldots u^p$. Show

$$dx^{i_1} \wedge \cdots \wedge dx^{i_p} = \frac{\partial(x^{i_1}, \ldots, x^{i_p})}{\partial(u^1 \ldots u^p)}\,du^1 \wedge \cdots \wedge du^p.$$

2. Let S be a two dimensional surface in $\mathbb{R}^3$ parametrized by $\vec{X}(u, v) = (x(u, v), y(u, v), z(u, v))$. The element of area on the surface is given by $d\vec{S} = \vec{X}_u \times \vec{X}_v\,du\,dv$. If $\vec{F}$ is a vector field, $\int\int_S \vec{F} \cdot d\vec{S} = \int\int_U F \cdot (X_u \times X_v)\,du\,dv$. Compute $F \cdot (X_u \times X_v)$ and show how to obtain Gauss' theorem from (5.1).

3. Derive Stokes' theorem

$$\int\int_S (\text{curl } \vec{V}) \cdot \hat{n}\,dS = \int_{\partial S} \vec{V} \cdot d\vec{x}$$

from (5.1), where S is a smooth surface in 3 dimensions.

19. Transformation Groups and Frame Invariance

Suppose we describe points in space relative to two coordinate systems (sometimes called *observers* or *frames*) labeled by $\mathcal{O}$ and $\mathcal{O}'$. A point P has coordinates x relative to $\mathcal{O}$ and x' relative to $\mathcal{O}'$ (see Fig. 5.1). Let S be some scalar physical quantity in space (such as the temperature) and suppose we represent $S(P)$ in each of the two reference frames. Relative to $\mathcal{O}$ the quantity S is given by a function $f(x)$, while relative to $\mathcal{O}'$ it is given by a different function $f'(x')$. Since S is an intrinsic quantity it is independent of the reference frame, so we must have

$$f(x) = f'(x'). \tag{5.2}$$

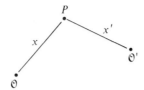

Figure 5.1

If φ represents the coordinate transformation $x' = \varphi(x)$ then from (5.2) we find that the functions f' and f are related by $f' = f \circ \varphi^{-1}$. Let us define the operation

$$\rho_\varphi f = f \circ \varphi^{-1}.$$

If we are considering not one but an entire group of coordinate transformations $\{\varphi\}$ then ρ_φ is a representation of that group by automorphisms of the algebra $C^\infty(M)$: $\rho_{\varphi_1 \varphi_2} = \rho_{\varphi_1} \rho_{\varphi_2}$.

Similarly, suppose a vector field V is represented in the two coordinate systems x and x' by

$$X = X^i(x)\frac{\partial}{\partial x^i} \quad \text{and} \quad X' = X'^i(x')\frac{\partial}{\partial x'^i}.$$

Since X and X' represent the same vector field V we require

$$X'f' = (Xf)'. \tag{5.3}$$

On the left side we have computed in the primed coordinates; while on the right side we first compute in the unprimed coordinates then transform the result to the frame $\mathcal{O}'$. The two results must be the same; hence (5.3).

Equation (5.3) leads directly to the transformation law for the components X' under a coordinate transformation $x' = \varphi(x)$ viz.

$$X'^i(\varphi(x))\frac{\partial x^j}{\partial x'^i} = X^j(x). \tag{5.4}$$

In the language of tensor analysis, the components X^i transform as a *contravariant* tensor. Note that the rule (5.4) is obtained directly from the chain rule:

$$X'^i(x')\frac{\partial}{\partial x'^i} = X'^i(\varphi(x))\frac{\partial x^j}{\partial x'^i}\frac{\partial}{\partial x^j} = X^j(x)\frac{\partial}{\partial x^j}.$$

The coordinate transformation $x' = \varphi(x)$ induced an action ρ_φ on $C^\infty(M)$ given by $\rho_\varphi f = f \circ \varphi^{-1}$. It also induces an action on $\mathcal{D}^1(M)$, the vector fields on M. Setting $X' = \rho_\varphi X$ we obtain from (5.3) the condition $(\rho_\varphi X)(\rho_\varphi f) = \rho_\varphi(Xf)$ hence

$$\rho_\varphi X = \rho_\varphi X \rho_{\varphi^{-1}}. \tag{5.5}$$

We confess to a slight ambiguity of notation here. On the right side of (5.5) $\rho_{\varphi^{-1}}$ acts on the function f, X acts on $\rho_{\varphi^{-1}}f$, and ρ_φ acts on the function $X\rho_{\varphi^{-1}}f$. On the left side $\rho_\varphi X$ denotes the action of ρ_φ on the vector field X. This ambiguity has its advantages. For example, it follows immediately from (5.5) that $\rho_{\varphi_1\varphi_2} = \rho_{\varphi_1}\rho_{\varphi_2}$, so ρ_φ is also a representation of any transformation group $\{\varphi\}$ on the vector fields $\mathscr{D}^1(M)$.

Similarly, an action of a transformation group on the one-forms $\Lambda_1(M)$ is obtained from the invariance condition

$$\omega'(X') = (\omega(X))'. \tag{5.6}$$

Setting $\omega' = \rho_\varphi\omega$ we find $(\rho_\varphi\omega) = \rho_\varphi\omega\rho_{\varphi^{-1}}$, $viz.$ $(\rho_\varphi\omega)(Y) = \rho_\varphi(\omega(\rho_{\varphi^{-1}}Y))$. As before one finds $\rho_{\varphi_1\varphi_2} = \rho_{\varphi_1}\rho_{\varphi_2}$. If $x' = \varphi(x)$ the chain rule leads directly to the transformation law for the components of a 1-form:

$$\omega_i'(x')\,dx'^i = \omega_i'(\varphi(x))\frac{\partial\varphi^i}{\partial x^j}\,dx^j = \omega_j(x)\,dx^j,$$

hence

$$\omega_i'(\varphi(x))\frac{\partial\varphi^i}{\partial x^j} = \omega_j(x).$$

The components ω_i transform as a $covariant$ tensor.

A transformation φ of a manifold M onto itself may be considered from either an "active" or "passive" viewpoint. In the passive viewpoint φ is regarded simply as a coordinate transformation: points on M do not move; one simply changes the way in which points, functions, tensors, etc. are represented. That is the view taken so far in our discussion. In the "active" view φ is regarded as a motion on M that moves points about. This is the point of view we take in thinking of the flow φ_t generated by a vector field. In this case, if, say φ moves p to q then ρ_φ moves functions, vectors and forms at p to q in the following sense. If U is a neighborhood of p and $V = \varphi(U)$ then ρ_φ carries $C^\infty(U)$ to $C^\infty(V)$. The reader may also show, by (5.5), that ρ_φ carries $\mathscr{D}^1(U)$ to $\mathscr{D}^1(V)$. (By $\mathscr{D}^1(U)$ we mean the vector fields in $\mathscr{D}^1(M)$ restricted to the neighborhood U.)

The action of the transformation φ on a general p-form is obtained from the invariance condition.

$$\omega'(X_1',\ldots,X_p') = (\omega(X_1,\ldots,X_p))'. \tag{5.7}$$

As before, we find that $\rho_\varphi\omega = \omega'$ defines a group action on the p-forms $\Lambda_p(M)$.

From (5.7) and the properties of the wedge product we easily find

$$\omega' \wedge \mu' = (\omega \wedge \mu)',$$

or equivalently, that $\rho_\varphi(\omega \wedge \mu) = \rho_\varphi\omega \wedge \rho_\varphi\mu$. This result is to be interpreted as a statement that the wedge product is invariant under arbitrary coordinate transformations. The same invariance is possessed by the exterior derivative and the Lie bracket. We have

Theorem 5.4. *For an arbitrary C^∞ coordinate transformation $x' = \varphi(x)$ on a manifold M we have*

$$\omega' \wedge \mu' = (\omega \wedge \mu)', \qquad d\omega' = (d\omega)', \qquad [X', Y'] = [X, Y]' \qquad (5.8)$$

or, equivalently,

$$\rho_\varphi \omega \wedge \mu = \rho_\varphi \omega \wedge \rho_\varphi \mu, \qquad d\rho_\varphi \omega = \rho_\varphi \, d\omega,$$

$$[\rho_\varphi X, \rho_\varphi Y] = \rho_\varphi [X, Y]. \qquad (5.8')$$

As a result of Theorem 5.4 we say that the operations $\wedge$, d and $[\,,\,]$ are *frame invariant*. We shall discuss this notion below at greater length.

PROOF. For $f \in C^\infty(M)$ we have $[X', Y']f' = X'Y'f' - Y'X'f' = X'(Yf)' - Y'(Xf)' = (XYf)' - (YXf)' = ([X, Y]f)' = [X, Y]'f'$. Since f is arbitrary we must have $[X', Y'] = [X, Y]'$.

Now let us prove that the exterior derivative is frame invariant.

First we prove the result for functions. We have $(df')(X') = X'f' = (Xf)' = (df(X))' = df'(X')$. Since X' is an arbitrary vector field we conclude that $df' = (df)'$. Next, for $\omega = df$ we have $d\omega' = d(df') = d(df') = d^2 f' = 0$; while $d(df)' = (d^2 f)' = 0$. Now suppose $\omega = g\,df$. Then $d\omega' = d(g'(df')) = d(g'\,df') = dg' \wedge df' = (dg \wedge df)' = (d(g\,df))' = (d\omega)'$. The theorem can be extended to forms of any order by showing that $d(\omega \wedge \mu)' = (d(\omega \wedge \mu))'$. We leave this last step to the reader. $\qquad\qquad\square$

That the Lie bracket and exterior derivative are "frame invariant" means that they have the same form in any coordinate system. Such is not the case, for example, for the operations of vector analysis such as the gradient, divergence, and curl. These operations are invariant under Euclidean motions of $\mathbb{R}^3$ but not under general coordinate transformations, as anyone who has calculated the divergence in spherical coordinates will quickly testify.

Let us pursue the notion of "frame invariance" a bit further. An expression Q (which may also be an operator) is said to be invariant under a coordinate transformation φ if $Q' = Q$, that is, if the representation of Q is the same in either coordinate system. Thus a function f is invariant if $f' = f$, that is, if $f(x') = f(x)$. For example, the function $x^2 + y^2$ is invariant under rotations about the origin: $x'^2 + y'^2 = x^2 + y^2$ if (x, y) and (x', y') are related by a rotation.

Similarly a differential form ω is invariant under the transformation φ if $\omega' = \omega$, i.e. if $\rho_\varphi \omega = \omega$. The two-form $dx \wedge dy$, for example, is invariant under rotations about the origin: if $x' = x\cos\theta + y\sin\theta$, $y' = -x\sin\theta + y\cos\theta$, then $dx' \wedge dy' = dx \wedge dy$.

The condition $\omega' = \omega$ means that the components of ω are the same in both coordinate systems. For a 1-form this means that

$$\omega_i(x')\,dx'^i = \omega_i(x)\,dx^i \qquad (5.9)$$

with the substitution $x' = \varphi(x)$ understood on the left. A more precise, but also more combersome statement, would be

$$\omega_i(\varphi(x))\frac{\partial \varphi^i}{\partial x^j} = \omega_j(x). \tag{5.9'}$$

The invariance of a p-form or a vector field under a coordinate transformation $x' = \varphi(x)$ is defined in analogous ways. A vector field X is invariant if $X' = X$, or $\rho_\varphi X = X$; a p-form is invariant if $\rho_\varphi \omega = \omega$.

In the same way, an operator T acting on $C^\infty(M)$ is invariant under a transformation φ if the representation of T is the same in both reference frames. Let $x' = \varphi(x)$ and let L' and L be the representations of T in the two coordinate systems. Then we have $L'u' = (Lu)'$. Putting $u' = \rho_\varphi u$ this condition says that $L' = \rho_\varphi L \rho_{\varphi^{-1}}$. We say T is invariant if $T' = T$; equivalently, T is invariant if $\rho_\varphi L = L \rho_\varphi$.

For example, the Laplacian $\Delta = \sum_{i=1}^{3} \frac{\partial^2}{\partial x^{i2}}$ is invariant under the Euclidean group of rigid motions. Suppose $x'^i = O_{ij}x^j + b_j$. Then $\frac{\partial x'^i}{\partial x^j} = O_{ij}$ and by the chain rule

$$\frac{\partial}{\partial x^i} = \frac{\partial x'^j}{\partial x^i}\frac{\partial}{\partial x'^j} = O_{ji}\frac{\partial}{\partial x'^j},$$

so

$$\Delta = \delta_{ij}\frac{\partial^2}{\partial x^i \partial x^j} = \delta_{ij}O_{mi}\frac{\partial}{\partial x'^m}O_{nj}\frac{\partial}{\partial x'^n}$$

$$= O_{mi}O_{ni}\frac{\partial^2}{\partial x'^m \partial x'^n} = \delta_{mn}\frac{\partial^2}{\partial x'^m \partial x'^n}$$

$$= \Delta'.$$

Thus the form of the Laplacian is the same in any Euclidean reference frame. We may say that the Laplacian is a Euclidean operator in the sense that it is invariant under Euclidean transformations. It is not, however, frame invariant; that is, it is not invariant under arbitrary coordinate transformations. (For example, compute the Laplacian in spherical coordinates.)

The operations grad, div, and curl are similarly Euclidean invariant but not frame invariant. Under a general coordinate transformation we have

$$\text{grad}' f' = (\text{grad } f)'$$
$$\text{div}' \vec{A}' = (\text{div } \vec{A})'$$
$$\text{curl}' \vec{A}' = (\text{curl } \vec{A})'.$$

For Euclidean coordinate transformations div $=$ div$'$, etc., but for a general change of frame, this is no longer true. As another example, the 2-form $dx \wedge dy$ in the plane is Euclidean invariant but not frame invariant, since in polar coordinates it is given by $r\, dr \wedge d\theta$.

These observations should help bring the content of Theorem 5.4 into better perspective. The operations $\wedge$, d, and $[\,,\,]$ are computed the same way in *any* (smooth) coordinate system. They are thus *frame invariant*.

The reader will recall that at the beginning of this chapter we showed that the Navier–Stokes equations are Euclidean invariant. If $u(x)$, $p(x)$ are the velocity and pressure in one reference frame, then $u'(x') = Ou(x)$ and $p'(x') = p(x)$ are the components in a second reference frame related to the first by a rigid motion. (That is, by an isometry of $\mathbb{R}^3$). The equations in the primed coordinates are exactly the same as they are in the unprimed coordinates.

We close this section with some remarks about our notation ρ_φ, which might be regarded as iconoclastic in some mathematical circles. For invertible transformations φ it coincides exactly with the action $(\varphi^{-1})^*$ on forms and φ_* on vector fields, a notation commonly found in modern texts on differential geometry. (See, for example, Singer and Thorpe, or Helgason.) We have concerned ourselves here with transformation groups $\{\varphi\}$ acting on a manifold M; hence φ is always invertible. (For non-invertible transformations φ_* and φ^* are still defined, whereas $(\varphi^{-1})^*$ is not.) We chose the notation ρ_φ to emphasize that the group action on the manifold M naturally induces a group action on the tensor spaces (vectors, forms, etc.) over M. In all cases,

$$\rho_{\varphi_1 \varphi_2} = \rho_{\varphi_1} \rho_{\varphi_2}$$

$$\rho_\varphi \omega \wedge \mu = \rho_\varphi \omega \wedge \rho_\varphi \mu.$$

More generally, $\rho_\varphi S \otimes T = \rho_\varphi S \otimes \rho_\varphi T$ where $\otimes$ is the tensor product (a notion we have not discussed here). Thus $\varphi \to \rho_\varphi$ is a representation of a transformation group $\mathfrak{G}$ by automorphisms of the bundle of all tensors on the manifold M. We shall make use of this fact in the next section.

If φ takes a point p to q then ρ_φ takes all vectors, forms, tensors based at p to vectors, forms, etc. based at q. By contrast, φ^* takes forms at q to forms at p and φ_* takes vectors at p to vectors at q.

EXERCISES

1. Show that $dx \wedge dt$ is invariant under the Lorentz transformation $x' = \gamma(x - vt)$, $t' = \gamma(t - vx)$, where $\gamma = (1 - v^2)^{-1/2}$. Show $dx \wedge dy \wedge dz \wedge dt$ is invariant under all proper ($\det \Lambda = 1$) Lorentz transformations.

2. Show the d'Alembertian $\Delta = \dfrac{\partial^2}{\partial t^2} - \Delta$ is invariant under the Poincaré group (translations plus Lorentz transformations of space-time).

3. How do the components $F_{\mu\nu}$ of the electromagnetic field transform under a Lorentz transformation?
   (Hint: Use $F_{\mu\nu}(x)\, dx^\mu \wedge dx^\nu = F'_{\mu\nu}(x')\, dx'^\mu \wedge dx'^\nu$ where $x' = \Lambda x$.)

4. If φ is a transformation on M and Ω is a p-dimensional submanifold of M, show that for any p-form ω

$$\int_{\Omega} \omega = \int_{\Omega'} \omega'$$

where $\Omega' = \varphi(\Omega)$ and $\omega' = \rho_{\varphi}\omega$. In other words, the integral of a p-form is frame invariant. If φ_t is a one-parameter group of transformations of M, show

$$\int_{\Omega_t} \omega = \int_{\Omega} \rho_{\varphi_{-t}}\omega$$

where $\Omega_t = \varphi_t(\Omega)$.

20. The Lie Derivative

Let X be a vector field on a manifold M, φ_t the flow generated by X, and put $\rho_t = \rho_{\varphi_{-t}}$, where ρ_{φ} is the group action discussed in §19. If φ_t carries m to $\varphi_t(m)$ then ρ_t pulls vectors and forms at $\varphi_t(m)$ back to m, as indicated in Fig. 5.2.

Since $\varphi_{t+s} = \varphi_t \circ \varphi_s$, we have $\rho_{t+s} = \rho_t \rho_s$. Furthermore ρ_t is a group of automorphisms of $C^{\infty}(M)$, of $\mathcal{D}^1(M)$, and of $\Lambda(M)$: that is

$$\rho_t(fg) = (\rho_t f)(\rho_t g) \qquad \text{for} \quad f, g \in C^{\infty}(M)$$

$$\rho_t[X, Y] = [\rho_t X, \rho_t Y] \qquad \text{for} \quad X, Y \in \mathcal{D}^1(M)$$

$$\rho_t \omega \wedge \mu = \rho_t \omega \wedge \rho_t \mu, \qquad \omega, \mu \in \Lambda(M).$$

Moreover, $\rho_t \, d\omega = d\rho_t \omega$.

Let L_X denote the infinitesimal generator of the one parameter group ρ_t. L_X is called the Lie derivative. By our remarks in §12 we know that L_X is a derivation on the algebra $\Lambda(M)$ and on the Lie algebra $\mathcal{D}^1(M)$. This fact is reflected in the following properties of L_X:

$$L_X fg = (L_X f)g + f(L_X g) \tag{5.10a}$$

$$L_X[Y, Z] = [L_X Y, Z] + [Y, L_X Z] \tag{5.10b}$$

$$L_X \omega \wedge \mu = L_X \omega \wedge \mu + \omega \wedge L_X \mu. \tag{5.10c}$$

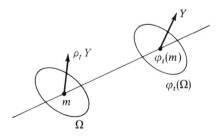

Figure 5.2

Furthermore, since $\rho_t d = d\rho_t$ we have

$$L_X \, d\omega = d L_X \omega. \tag{5.10d}$$

We already saw in §17 that $L_X f = X f$ for $f \in C^\infty(M)$. We claim the Lie derivative on $\mathscr{D}^1(M)$ is given by

$$L_X Y = \operatorname{ad} X(Y) = [X, Y]. \tag{5.11}$$

In fact, by (5.5) $\rho_t Y = \rho_t Y \rho_{-t}$, so

$$\frac{d}{dt}(\rho_t Y)f\big|_{t=0} = \frac{d}{dt}\rho_t(Yf)\big|_{t=0} + Y\frac{d}{dt}\rho_{-t}f\big|_{t=0} = XYf - YXf = [X, Y]f.$$

An intrinsic (coordinate-free) expression for the Lie derivative on forms can also be given (see, for example, Sternberg, p. 102) but requires somewhat more machinery then we have developed here. It will be sufficient, for our purposes, to compute the action of L_X on forms in local coordinates. For this we need only the linearity of L_X and properties (5.10c, d). If $X = X^j(\partial/\partial x^j)$ and $\omega = \omega_l \, dx^l$ then

$$\begin{aligned}
L_X \omega &= (L_X \omega_l)\, dx^l + \omega_l L_X \, dx^l \\
&= (X\omega_l)\, dx^l + \omega_l \, dX x^l \\
&= X\omega_l \, dx^l + \omega_l \, dX^l \\
&= \left(X^j \frac{\partial \omega_l}{\partial x^j} + \omega_j \frac{\partial X^j}{\partial x^l} \right) dx^l.
\end{aligned}$$

The Lie derivative of any p-form can be computed simply by applying the rules of linearity and using (5.10c) and (5.10d).

We close this section with the transport theorem:

Theorem 5.5. *Let φ_t be a flow on a manifold M, let ω be a p-form, and let Ω be a smooth p-dimensional submanifold. Then, with $\Omega_t = \varphi_t(\Omega)$,*

$$\frac{d}{dt} \int_{\Omega_t} \omega = \int_{\Omega_t} L_X \omega.$$

PROOF. By changing variables we can bring the integral of ω over the surface Ω_t down to an integral over the fixed surface Ω, namely

$$\int_{\Omega_t} \omega = \int_{\Omega} \rho_t \omega.$$

(This equivalence could also be derived from the principle of frame invariance of the integral: see exercise 4, §19.) The derivative of the right hand side is

$$\frac{d}{dt} \int_{\Omega} \rho_t \omega = \int_{\Omega} \frac{d}{dt} \rho_t \omega = \int_{\Omega} \rho_t L_X \omega.$$

Transforming back, we get

$$\frac{d}{dt}\int_{\Omega_t}\omega = \int_{\Omega}\rho_t L_X\omega = \int_{\Omega_t} L_X\omega. \qquad \square$$

The transport theorem is the source of numerous conservation laws in mathematical physics (see the exercises below).

EXERCISES

1. If ω is a p-form which depends on t, show that

$$\frac{d}{dt}\int_{\Omega_t}\omega(t) = \int_{\Omega_t}\left(\frac{\partial\omega}{\partial t} + L_X\omega\right)$$

where Ω_t moves with the flow generated by X.

2. Let $\rho(x, t)$ be the density of a fluid and let $v(x)$ be the velocity of the fluid in a region G of $\mathbb{R}^n$ ($x \in \mathbb{R}^n$ and $v(x)$ is a vector field in $\mathbb{R}^n$). By conservation of mass the integral

$$\int_{\Omega(t)}\rho(x, t)\,dx$$

is constant if $\Omega(t)$ is any n-dimensional region which moves with the flow. Use the transport theorem to show that

$$\rho_t + \sum_{i=1}^{n}\frac{\partial}{\partial x^i}(\rho v^i) = 0$$

3. The Euler equations for an inviscid incompressible fluid in Cartesian coordinates are

$$\frac{\partial u_i}{\partial t} + u^j\frac{\partial u_i}{\partial x^j} = -\frac{\partial p}{\partial x^i}.$$

In Cartesian coordinates we have $u_i = u^i$. Prove Kelvin's theorem of hydrodynamics:

$$\frac{d}{dt}\int_{C(t)} u_i\,dx^i = 0$$

where $C(t)$ is a closed curve which moves with the fluid.

4. A vector field Y is invariant under the flow generated by X if $\rho_t Y = Y$. Let $X = y\frac{\partial}{\partial x} - x\frac{\partial}{\partial y}$ and find the invariant vector fields. Show they form a Lie algebra. Formulate a definition of invariant one-forms and find them for X. 2-forms?

5. Consider a Hamiltonian flow given in local coordinates by

$$\dot{p}_i = \frac{\partial H}{\partial q_i} \qquad \dot{q}_i = -\frac{\partial H}{\partial p_i}.$$

Denote the corresponding vector field by X_H. Let $\Omega = \sum_{i=1}^{n} dp_i \wedge dq_i$, and show that $L_{X_H}\Omega = 0$. What is $\Omega \wedge \Omega \wedge \cdots \wedge \Omega$? Show $L_{X_H}\Omega \wedge \cdots \wedge \Omega = 0$.

CHAPTER 6

Symmetry Groups of Differential Equations

21. Group Actions on Jet Bundles

Consider an ordinary differential equation $f(x, u, u') = 0$ and assume that the set $f(x, u, p) = 0$ is a smooth submanifold of $\mathbb{R}^3$. A solution of the differential equation is a curve $(x, u(x), p(x))$ on the surface $f = 0$ such that $p(x) = u'(x)$. We wish to know what transformations of the x-u plane leave the set of solutions invariant.

Let a one-parameter group of transformations in the x-u plane be given by

$$\tilde{x} = X(x, u, \varepsilon) \qquad \tilde{u} = U(x, u, \varepsilon). \tag{6.1}$$

Under this transformation the graph of a function $u = u(x)$ (given by $\Gamma = \{(x, u) | u = u(x)\}$ is transformed (at least locally) into a new graph $\tilde{\Gamma}$. The transformation (6.1) thus carries the original function $u(x)$ into a new one $\tilde{u}(\tilde{x})$; and moreover a point $(x, u(x), (du/dx))$ is carried into $(\tilde{x}, \tilde{u}(\tilde{x}), (d\tilde{u}/d\tilde{x}))$.

The new function $\tilde{u}(\tilde{x})$ is represented parametrically by

$$\tilde{x} = X(x, u(x), \varepsilon) \qquad \tilde{u} = U(x, u(x), \varepsilon). \tag{6.2}$$

For $\varepsilon = 0$ $X(x, u, 0) = x$ so (6.2) may be inverted for x as a function of $\tilde{x}$ for small ε, and the result may then be substituted into the equation for $\tilde{u}$ to obtain $\tilde{u}(\tilde{x})$.

The computation of $\tilde{u}(\tilde{x})$ and of $\tilde{p} = d\tilde{u}/d\tilde{x}$ is in general quite complicated; but let us carry out the computation in a simple case to see what happens. Consider the group of rotations

$$\tilde{x} = x \cos \varepsilon - u \sin \varepsilon$$
$$\tilde{u} = x \sin \varepsilon + u \cos \varepsilon \tag{6.3}$$

acting in the x-u plane. The function $u(x) = px + b$ is carried into a new straight line by the rotation, and one may easily calculate that.

$$x = \frac{\tilde{x} + b \sin \varepsilon}{\cos \varepsilon - p \sin \varepsilon}$$

$$\tilde{u}(\tilde{x}) = \frac{\tilde{x}(\sin \varepsilon + p \cos \varepsilon) + b}{(\cos \varepsilon - p \sin \varepsilon)}$$

for sufficiently small ε. The action induced on (x, u, p) space is given by

$$(x, u, p) \to \left(x \cos \varepsilon - u \sin \varepsilon, x \sin \varepsilon + u \cos \varepsilon, \frac{\sin \varepsilon + p \cos \varepsilon}{\cos \varepsilon - p \sin \varepsilon} \right)$$

$$= (\tilde{x}, \tilde{u}, \tilde{p}) = \Phi_\varepsilon^{(1)}(x, u, p).$$

We leave it to the reader as an exercise to show that the $\Phi_\varepsilon^{(1)}$ form a one-parameter transformation group on $\mathbb{R}^3$. Thus the group action (6.3) on the plane has induced a group action $\Phi_\varepsilon^{(1)}$ on $\mathbb{R}^3$, called the *first prolongation*. The infinitesimal generator of the action (6.3) is $\alpha = -u(\partial/\partial x) + x(\partial/\partial u)$. The reader may easily show that the infinitesimal generator of $\Phi_\varepsilon^{(1)}$ is

$$pr^{(1)}\alpha = \alpha + (1 + p^2)\frac{\partial}{\partial p}.$$

The vector field $pr^{(1)}\alpha$ is called the *first prolongation* of α.

We have obtained the prolongation of (6.3) by direct calculation, but the general group action (6.2) also has a prolongation. Let Φ_g be an action in the x-u plane. Since $\Phi_{g_1 g_2} = \Phi_{g_1} \circ \Phi_{g_2}$ we have $\tilde{\Gamma}_{g_1 g_2} = \Phi_{g_1 g_2}(\Gamma) = \Phi_{g_1}(\Phi_{g_2}(\Gamma))$. Therefore the group composition property must hold for the slope of Γ at every point, and there is a natural prolongation of any group action on $\mathbb{R}^2$ to $\mathbb{R}^3$. Moreover, we can extend this argument to cover derivatives of all orders of the initial function $u(x)$.

The kth prolongation of the action Φ_ε on $\mathbb{R}^2$ is the action induced on kth order Taylor polynomials as follows. Let $X \times U$ be the Cartesian product of pairs (x, u) and let $\mathcal{J}_k$ be the vector space $X \times U \times \mathbb{R}^k$ with coordinates labelled $(x, u, u^1, \ldots, u^k)$. $\mathcal{J}_k$ is called the kth order *jet bundle*. Given a point $(\bar{x}, \bar{u}, \bar{u}^1, \ldots, \bar{u}^k)$ in $\mathcal{J}_k$ we construct the polynomial

$$f(x) = \sum_{j=0}^{k} \frac{\bar{u}^j (x - \bar{x})^j}{j!}$$

where $\bar{u}^0 = \bar{u}$. A smooth transformation Φ on the base space $X \times U$ will carry $f(x)$ to a new function $\tilde{f}(\tilde{x})$ as described in (6.2). This procedure leads in a natural way to a transformation $x \to \tilde{x}$, $\bar{u}^j \to \tilde{\bar{u}}^j$ on the space $\mathcal{J}_k$. Furthermore, an action Φ_g on $X \times U$ induces an action $\Phi_g^{(k)}$ on $\mathcal{J}_k$, called the kth *prolongation*.

In particular, if $\mathfrak{G}$ is a Lie group of transformations acting on $X \times U$, then transformations sufficiently close to the identity will preserve graphs and will induce a transformation on $\mathcal{J}_k$. Since the group composition is preserved by the prolongation the infinitesimal generators $pr^{(k)}\alpha$ of the kth prolongation will form a representation of the Lie algebra $\mathfrak{g}$, i.e.

$$[pr^{(k)}(\alpha), pr^{(k)}(\beta)] = pr^{(k)}[\alpha, \beta].$$

There are no difficulties in extending all these concepts to several variables. We write $x = (x_1, \ldots, x_n)$ and $u = (u^1, \ldots, u^m)$ and take the base space to be $X \times U = (x_1, \ldots, x_n, u^1, \ldots, u^m)$. For a multi-index $J = (j_1, \ldots, j_n)$ where $j_1, \ldots, j_n$ are non-negative integers, let $|J| = j_1 + \cdots + j_n$ and let

$$\partial_J = \frac{\partial^{|J|}}{\partial x_1^{j_1} \cdots \partial x_n^{j_n}}.$$

Let $\mathcal{J}$ be the linear space whose coordinates are labelled by $x_1, \ldots, x_n, u^1, \ldots, u^m, u^l_J$ where $l = 1, \ldots, m$ and $|J| \leq k$. Every vector valued function $u = u(x) = (u^1(x), \ldots, u^m(x))$ defines a *section* in $\mathcal{J}_k$ given by $(x_i, u^l(x), u^l_J(x))$ where $u^l_J(x) = \partial_J u^l(x)$. The notation u^l_J denotes the coordinates in $\mathcal{J}_k$ and $\partial_J u^l$ denotes the derivatives of a function $u^l(x)$.

We calculate the kth prolongation of a vector field α on $X \times U$ in several steps. Define on the spaces $\mathcal{J}_k$ the formal vector field

$$D_i = \frac{\partial}{\partial x_i} + \sum_{l,J} u^l_{J_i} \frac{\partial}{\partial u^l_J}$$

where $J_i = (j_1, \ldots, j_i + 1, \ldots, j_n)$. D_i acts as the total derivative d/dx_i on sections. That is, if $(x, u(x), u_J(x))$ is a section then $D_i f(x, u(x), u_J(x)) = (d/dx_i) f(x, u(x), u_J(x))$. Now consider a general one parameter transformation group

$$\tilde{x}_i = X^i(x, u, \varepsilon), \qquad \tilde{u}^j = U^j(x, u, \varepsilon) \tag{6.4}$$

where $1 \leq i \leq n$ and $1 \leq j \leq m$. Let the infinitesimal generator of this one parameter group be given by

$$\alpha = \xi^i \frac{\partial}{\partial x_i} + \varphi^j \frac{\partial}{\partial u_j} \tag{6.5}$$

where $\xi^i = \xi^i(x, u)$ and $\varphi^j = \varphi^j(x, u)$. If $\delta = d/d\varepsilon|_{\varepsilon=0}$ then $\delta X^i = \xi^i$ and $\delta U^j = \varphi^j$.

Theorem 6.1. *The first prolongation of α is given by*

$$pr^{(1)}\alpha = \alpha + \sum_{l,i} \varphi^l_i \frac{\partial}{\partial u^l_i} \tag{6.6}$$

where

$$\varphi^l_i = D_i \varphi^l - u^l_m D_i \xi^m. \tag{6.7}$$

PROOF. A section $(x, u(x), u_j(x))$ is transformed into $(\tilde{x}, \tilde{u}(\tilde{x}), \tilde{u}_j(\tilde{x}))$ where $u_j(x) = \partial u/\partial x_j$ and $\tilde{u}_j(\tilde{x}) = \partial \tilde{u}(\tilde{x})/\partial \tilde{x}_j$.

From (6.4)

$$\frac{\partial \tilde{x}_m}{\partial x_i} = D_i X^m \qquad \frac{\partial u^l}{\partial x_i} = D_i u^l$$

when the transformations are restricted to sections. Therefore

$$D_i U^l = \frac{\partial \tilde{u}^l}{\partial x_i} = \frac{\partial \tilde{u}^l}{\partial \tilde{x}_m}\frac{\partial \tilde{x}_m}{\partial x_i} = \frac{\partial \tilde{u}^l}{\partial \tilde{x}_m} D_i X^m.$$

Putting $D_x X = \| D_i X^j \|$, $D_x U = \| D_i U^l \|$ and $\tilde{P} = \| \partial \tilde{u}^l / \partial \tilde{x}_i \|$ we have $\tilde{P}^l_m D_i X^m = D_i U^l$, or $\tilde{P}(D_x X) = D_x U$.

The functions φ^l_i in (6.7) are given by $\varphi^l_i = \delta \tilde{u}^l_i$ so we must compute $\delta \tilde{P}$. Since $\tilde{P} = P$ and $D_x X = I$ at $\varepsilon = 0$,

$$\delta(D_x U) = \delta \tilde{P} + P \delta(D_x X);$$

and, since δ commutes with D_x, we get

$$\delta \tilde{P} = D_x \delta U - P D_x(\delta X) = D_x \varphi - P D_x \xi.$$

This is the relationship (6.7). $\qquad\square$

The higher order prolongations are calculated by a recursion relation. Having found $\alpha_k = pr^{(k)}(\alpha)$ we compute α_{k+1} as $\alpha_{k+1} = pr^{(1)}(\alpha_k)$. Let us write

$$pr^{(k)}(\alpha) = \alpha + \sum_{l, |J| \le k} \varphi^l_J \frac{\partial}{\partial u^l_J}. \qquad (6.8)$$

Knowing the functions φ^l_J for $|J| \le k$ we apply $pr^{(1)}(\alpha_k)$ to the coordinates u^p_J to obtain $pr^{(k+1)}(\alpha)$. Replacing u^l_i by u^l_J and φ^l by φ^l_J in (6.7) we get

$$\varphi^l_{J_i} = D_i \varphi^l_J - u^l_{J_m} D_i \xi^m \qquad (6.9)$$

where $J_i = (j_1, \ldots, j_i + 1, \ldots, j_n)$. Iterating this recursion relation we get a general form for $pr^{(k)}(\alpha)$:

Theorem 6.2. *The functions φ^l_J in (6.8) are given by*

$$\varphi^l_J = D_J(\varphi^l - \xi^m u^l_m) + (D_m u^l_J)\xi^m. \qquad (6.10)$$

PROOF. Substituting (6.10) into the recursion relation (6.9) we get

$$\begin{aligned}
\varphi^l_{J_i} &= D_i \varphi^l_J - u^l_{J_m} D_i \xi^m \\
&= D_i(D_J(\varphi^l - \xi^m u^l_m) + u^l_{J_m}\xi^m) - u^l_{J_m} D_i \xi^m \\
&= D_{J_i}(\varphi^l - \xi^m u^l_m) + (D_i u^l_{J_m})\xi^m \\
&= D_{J_i}(\varphi^l - \xi^m u^l_m) + (D_m u^l_{J_i})\xi^m
\end{aligned}$$

which is (6.10) with J replaced by J_i. We used the facts that $u^l_{J_m} = D_m u^l_J$ and that $[D_i, D_m] = 0$; the proof that D_i and D_m commute is left as an exercise. $\qquad\square$

22. Infinitesimal Symmetries of Differential Equations

Let $\Delta(x, u, u_J)$ be a smooth mapping from $\mathscr{J}_k$ to R^q such that the set of x, u, u_J in $\mathscr{J}_k$ for which $\Delta = 0$ forms a smooth submanifold of $\mathscr{J}_k$. The set $\Delta = 0$ will be a smooth manifold of codimension q if the Jacobian

$$\frac{\partial(\Delta^1, \dots, \Delta^q)}{\partial(x, u^l, u_J^l)}$$

is of rank q everywhere on $\Delta = 0$. We regard $\Delta(x, u, u_J) = 0$ as a system of partial differential equations; by a solution we mean a *section* $(x, u(x), u_J(x))$ such that $\Delta(x, u(x), u_J(x)) = 0$ in some open domain of X.

Let U be a domain in $\mathbb{R}^n$ and let $F: U \to \mathbb{R}$. Let Φ_g be the action of a (local) Lie group of diffeomorphisms acting on U. F is invariant if $F(\Phi_g(x)) = F(x)$ for $x \in U$ and g sufficiently close to the identity so that $F \circ \Phi_g$ makes sense. For example functions of $\rho = \sqrt{x^2 + y^2 + z^2}$ are invariant under rotations in $\mathbb{R}^3$; functions $f(r, z)$, where $r = \sqrt{x^2 + y^2}$, are invariant under rotations about the z-axis.

Lemma 6.3. *A function F is invariant with respect to a Lie group $\mathfrak{G}$ iff $XF = 0$ for all X in the Lie algebra of $\mathfrak{G}$.*

PROOF. If F is invariant then $F(\Phi_t(x)) = F(x)$ for every one-parameter subgroup. Since $(d/dt)F(\Phi_t(x))|_{t=0} = XF$ where X is the infinitesimal generator of Φ_t we have $XF = 0$ for all X. Conversely, if $XF = 0$ then $(d/dt)F(\Phi_t(x)) = (\partial F/\partial x^i)(dx^i/dt) = XF = 0$ everywhere, so F is constant along trajectories of the group action generated by X. If $XF = 0$ for all X in the Lie algebra then F is invariant under every one parameter subgroup and therefore under the group itself. $\square$

A mapping $F: \mathbb{R}^n \to \mathbb{R}^m$ is invariant if each of its components is invariant. A submanifold $\mathscr{M}$ of $\mathbb{R}^n$ is invariant under a local group action Φ_g if $x \in \mathscr{M}$ implies $\Phi_g(x) \in \mathscr{M}$ for g sufficiently close to the identity. If F is invariant then obviously all its level surfaces are invariant. On the other hand it is possible that a particular level surface is invariant even though F is not. F is invariant if and only if all its level surfaces are invariant. We prove

Lemma 6.4. *Let U be a domain in $\mathbb{R}^n$, let $F: U \to \mathbb{R}^m$ and let $\mathscr{M}$ be a particular level surface of F, namely, $\mathscr{M} = \{x: F(x) = c, x \in U, c \in \mathbb{R}^m\}$. Suppose that $\|\partial F_i/\partial x_j\|$ is of rank m everywhere on $\mathscr{M}$. Then $\mathscr{M}$ is invariant under a group action Φ_g if and only if $XF = 0$ everywhere on $\mathscr{M}$ for all infinitesimal generators of the action.*

Note that we require $XF = 0$ only on $\mathscr{M}$ and not everywhere, so we are not requiring that F be invariant. This subtle point will be important in our calculations of symmetry groups of differential equations. By $XF = 0$ we mean $XF^i = 0$ for each component of F.

PROOF. If $\mathcal{M}$ is invariant and $x \in \mathcal{M}$ then $\Phi_t(x) \in \mathcal{M}$ for every one-parameter subgroup. Hence $F(\Phi_t(x)) = F(x)$ identically in t, and $XF = 0$ whenever $x \in \mathcal{M}$.

To prove the converse, we make the change of coordinates

$$\tilde{x}_j = F^j(x_1, \ldots, x_n) \qquad j = 1, \ldots, m$$

$$\tilde{x}_j = x_j \qquad\qquad j = m+1, \ldots, n.$$

Since $\| \partial F^i / \partial x_j \|$ is always of rank m we may suppose the variables $x_1, \ldots, x_n$ have been ordered in such a way that $\| \partial \tilde{x}_i / \partial x_j \|$ is invertible, so that our coordinate transformation is non-singular in a neighborhood of $\mathcal{M}$. In the coordinates $\tilde{x}_j$ the manifold $\mathcal{M}$ is given by $\mathcal{M} = \{ \tilde{x}_j = c^j; j = 1, \ldots, m \}$, and $F(\tilde{x}) = \{ \tilde{x}_1, \ldots, \tilde{x}_m \}$. The condition $XF = 0$ is

$$\tilde{X}^i \frac{\partial F^j}{\partial \tilde{x}_i} = \tilde{X}^i \delta_i^j = \tilde{X}^j = 0, \qquad j = 1, \ldots, m.$$

The equations generating the flow are

$$\frac{d\tilde{x}_i}{dt} = \tilde{X}^i \qquad i = 1, \ldots, n,$$

so $d\tilde{x}_i/dt = 0$ for $i = 1, \ldots, m$ and the manifold $\mathcal{M} = \{ \tilde{x}_j = c^j, j = 1, \ldots, m \}$ is therefore invariant. $\qquad\qquad\square$

Theorem 6.5. *Let $\mathfrak{G}$ be a local group acting on $X \times U$ by*

$$\tilde{x} = X(x, u, g), \qquad \tilde{u} = U(x, u, g) \tag{6.11}$$

and let $\Delta(x, u, u_j)$ be a system of partial differential equations which is invariant under the prolongation of the action $\Phi_g^{(k)}$ to $\mathcal{J}_k$. Then the action (6.11) preserves solutions of $\Delta = 0$; that is, if $u(x)$ is a solution so is $\tilde{u}(\tilde{x})$. A necessary and sufficient condition that the equations be invariant with respect to $\Phi_g^{(k)}$ is that

$$pr^{(k)}(\alpha)\Delta = 0 \quad whenever \quad \Delta = 0 \tag{6.12}$$

for every infinitesimal generator of the group action (6.11).

PROOF. If $u(x)$ is a solution then $\Delta(x, u(x), u_J(x)) = 0$ for x in an open domain of X. Since $\Phi_g^{(k)}$ preserves sections, the section $(x, u(x), u_J(x))$ goes into $(\tilde{x}, \tilde{u}(\tilde{x}), \tilde{u}_J(\tilde{x}))$ and the invariance of the submanifold $\Delta = 0$ under $\Phi_g^{(k)}$ means that $\Delta(\tilde{x}, \tilde{u}(\tilde{x}), \tilde{u}_J(\tilde{x})) = 0$ whenever $\Delta(x, u(x), u_J(x)) = 0$. The necessity and sufficiency of (6.12) follows from Lemma 6.4. $\qquad\square$

The condition (6.9) together with the formula (6.7) for the prolongation gives us a way to calculate the generators α of the symmetry group of a system of differential equations.

Example 6.6. Symmetries of $\Delta u = 0$ in $\mathbb{R}^n$. The equations for the symmetries are obtained from the condition $pr^{(2)}\alpha(\Delta u) = 0$ when $\Delta u = 0$:

$$\sum_{j=1}^{n} \varphi_{(jj)} = 0 \quad \text{when} \quad \sum_{j=1}^{n} u_{jj} = 0.$$

Here we use the notation $\varphi_{(jj)}$ for φ_J and u_{jj} for $\partial_J u$ when $J = (0, \ldots, \overset{j}{2}, \ldots, 0)$. Now

$$\varphi_{(jj)} = D_j^2(\varphi - \xi^i u_i) + \xi^i u_{jji}$$

$$= \varphi_{jj} + \varphi_{ju} u_j + \varphi_u u_{jj} + (\varphi_{uj} + \varphi_{uu} u_j)u_j - \xi^i u_{ijj} - 2u_{ij}(\xi_j^i + \xi_u^i u_j)$$

$$- u_i(\xi_{jj}^i + \xi_{ju}^i u_j + \xi_u^i u_{jj} + \xi_{uj}^i u_j + \xi_{uu}^i u_j u_j) + \xi^i u_{jji}.$$

Forming the sum we get

$$\sum_{j=1}^{n} \{\varphi_{jj} + u_j(2\varphi_{uj} - \Delta_x \xi^j)\} + \varphi_u \Delta u$$

$$- \sum_{i,j=1}^{n} u_{ij}(\xi_j^i + \xi_i^j) + u_i u_j(\varphi_{uu}\delta_{ij} - (\xi_{uj}^i + \xi_{ui}^j))$$

$$+ \sum_{i,j,k} u_i u_{jk}(-2\xi_u^j \delta_{ki} - \xi_u^i \delta_{kj}) - \sum_{i,j} u_i u_j^2 \xi_{uu} = 0$$

$$\text{when} \quad \sum_{j=1}^{n} u_{jj} = 0.$$

Now the variables u_i, u_{ij}, etc. are independent variables so we regard this as a polynomial in independent variables u_j, u_{ij} which must vanish identically on $\mathcal{J}_k$. Setting each of the coefficients equal to zero we get

(i) $\Delta_x \varphi(x, u) = 0 \qquad \left(\sum_{j=1}^{n} \varphi_{jj} = 0 \right).$

(ii) $2\varphi_{uj} - \Delta_x \xi^j = 0.$

(iii) $\xi_j^i + \xi_i^j = 0 \qquad i \neq j.$

(iv) $\xi_1^1 = \xi_2^2 = \cdots = \xi_n^n.$

(v) $\varphi_{uu}\delta_{ij} = (\xi_{uj}^i + \xi_{ui}^j).$

(vi) $2\xi_u^j \delta_{ki} + \xi_u^i \delta_{kj} = 0.$

(vii) $\xi_{uu}^i = 0.$

Condition (iv) is obtained as follows. We have

$$\sum_{i,j} u_{ij}(\xi_j^i + \xi_i^j) = \sum_{i \neq j} u_{ij}(\xi_j^i + \xi_i^j) + 2 \sum_{i=1}^{n} u_{ii}(\xi_i^i).$$

Equation (iii) follows from the fact that the u_{ij} are independent. On the manifold $\Delta u = 0$, we have $u_{11} = -u_{22} - \cdots - u_{uu}$; so

$$\sum_{i=1}^{n} u_{ii}\xi_i^i = \sum_{i=2}^{n} u_{ii}(\xi_i^i - \xi_1^1).$$

Since the variables u_{ii} are independent for $i = 2, \ldots, n$ we may conclude that $\xi_i^i = \xi_1^1$ for $i = 2, \ldots, n$.

We do not set $\varphi_u = 0$ since we are on the manifold $\Delta u = 0$. We now turn to the solutions of the partial differential equations (i)–(vii) for the infinitesimal generators of the symmetry group. Equation (vi) implies that $3\xi_u^i = 0$ (when $i = j = k$) so that $\xi^i = \xi^i(x)$. Then (v) shows that $\varphi = A(x)u + B(x)$. From (i) we see that $\Delta A = \Delta B = 0$. From (ii) and (iii) we have $2A_j = \Delta\xi^j$ and $2A_{jk} = \Delta\xi_k^j = -\Delta\xi_j^k = -2A_{kj}$; hence $A_{kj} = 0$ for $j \neq k$. In the same way, from (ii) and (iv) we find that $A_{11} = \cdots = A_{nn}$; but then $A_{ii} = 0$ since we already found that $\Delta A = 0$. So A is a linear function,

$$A(x) = A_0 + \sum_{i=1}^{n} A_i x_i.$$

Putting $\Psi = \xi_1^1 = \cdots = \xi_n^n = \Psi(x_1 \ldots x_n)$, we find that for $j \neq k$

$$0 = 2A_{jk} = \Delta\xi_k^j = \sum_{i=1}^{n} \xi_{kii}^j = -\sum_{i=1}^{n} \xi_{jki}^i = -n\Psi_{jk}.$$

Therefore

$$\Psi = \sum_{j=1}^{n} \psi_j(x_j).$$

Now

$$\psi_1''(x_1) = \Psi_{11} = \xi_{111}^1 = \xi_{211}^2 = \xi_{121}^2 = -\xi_{221}^1$$

$$= -\xi_{122}^1 = -\Psi_{22} = -\psi_2''(x_2).$$

Therefore

$$\psi_1''(x_1) + \psi_2''(x_2) = 0$$

and the two functions must be constant. The same is true of all the other functions, so

$$\Psi = a_1 x_1 + \cdots + a_n x_n + a_0$$

where the a_i are constants. We do not suppose all the constants a_i, A_i to be independent. Now $\xi^1, \ldots, \xi^n$ may be obtained by integrating Ψ so we get

$$\xi^1 = \frac{a_1}{2}x_1^2 + x_1(a_2 x_2 + \cdots + a_n x_n) + a_0 x_1 + C^1(x_2 \ldots x_n)$$

$$\xi^2 = \frac{a_2}{2}x_2^2 + x_2(a_1 x_1 + a_3 x_3 + \cdots + a_n x_n) + a_0 x_2 + C^2(x_1, x_3 \ldots x_n)$$

$$\vdots$$

where

$$C^1, C^2, \ldots \text{ are constants of integration.}$$

We find from (iii) that $\xi_{jk}^1 = 0$ for $j \neq k \neq 1$, so we must have

$$C^1(x_2 \ldots x_n) = C_2^1(x_2) + \cdots + C_n^1(x_n)$$
$$C^2(x_1, x_3, \ldots x_n) = C_1^2(x_1) + \cdots + C_n^2(x_n).$$

Therefore

$$\xi^l = \frac{a_l}{2} x_l^2 + x_l \sum_{l \neq j} a_j x_j + a_0 x_l + \sum_{j \neq l} C_j^l(x_j).$$

Now

$$\xi_j^i + \xi_i^j = a_i x_j + a_j x_i + C_j^{\prime i}(x_j) + C_i^{\prime j}(x_i) = 0,$$

so

$$C_j^{\prime i}(x_j) + a_i x_j = \gamma_j^i$$

where $\gamma_j^i + \gamma_i^j = 0, i \neq j$; and

$$C_j^i(x_j) = \frac{-a_i}{2} x_j^2 + \gamma_j^i x_j + \tau_j^i.$$

We therefore find

$$\xi^l(x^1, \ldots, x^n) = \frac{a_l}{2}(2x_l^2 - x_1^2 - \cdots - x_i^2) + \sum_{j \neq l} a_j x_j x_l + \gamma_j^l x_j + a_0 x^l + \tau^l$$

where $\gamma_j^l + \gamma_l^j = 0$. To find the A_i we use $2A_l = \Delta \xi^l = a_l(2 - n)$. Then

$$\varphi(x, u) = \sum_{l=1}^n \left(\frac{2 - n}{2}\right) a_l x_l u + A_0 u + B(x)$$

where $\Delta B = 0$.

Each of the constants above is associated with an independent infinitesimal symmetry, as follows

$$\tau^l: \quad \frac{\partial}{\partial x^l} \qquad \text{translations}$$

$$\gamma_j^l: \quad x_j \frac{\partial}{\partial x^l} - x_l \frac{\partial}{\partial x^j} \qquad \text{rotations}$$

$$a_0: \quad x^l \frac{\partial}{\partial x^l} \qquad \text{dilations}$$

(6.13)

$$A_0: \quad u \frac{\partial}{\partial u} \qquad u \to \lambda u \text{ (scalar multiplication)}$$

$$B: \quad B(x) \frac{\partial}{\partial u} \qquad u \to u + \lambda B(x), \qquad \Delta B = 0$$

$$a_k: \quad -\frac{1}{2} r^2 \frac{\partial}{\partial x^k} + x_k \left(\sum_j^n x_j \frac{\partial}{\partial x_j}\right) + \left(\frac{n - 2}{2}\right) x_k u \frac{\partial}{\partial u}.$$

The last vector fields generate the conformal transformations in $\mathbb{R}^n$.

23. Symmetries and Conservation Laws

Consider a functional

$$\int\int_\Omega \omega(x, u(x), \partial_J u(x)) \, dx. \tag{6.14}$$

Under the one parameter transformation group (6.4) the integral goes into

$$\int\int_{\tilde{\Omega}} \omega(\tilde{x}, \tilde{u}(\tilde{x}), \tilde{\partial}_J \tilde{u}(\tilde{x})) \, d\tilde{x}.$$

By the transport theorem (Theorem 5.5)

$$\frac{d}{d\varepsilon} \int\int_{\tilde{\Omega}} \omega(\tilde{x}, \tilde{u}, \tilde{\partial}_J \tilde{u}) \, d\tilde{x} = \int\int_{\tilde{\Omega}} L_\alpha(\tilde{\omega} \, d\tilde{x})$$

where L_α is the Lie derivative on the n form $\tilde{\omega} \, d\tilde{x}$. The integral is invariant under this transformation group if $L_\alpha(\tilde{\omega} \, d\tilde{x}) = 0$. We calculate this at the identity

$$L_\alpha \omega \, d\tilde{x} = (L_\alpha \omega) \, dx + \omega L_\alpha \, d\tilde{x} = pr^{(k)}\alpha(\omega) \, dx + \omega L_\alpha \, d\tilde{x}$$

since the Lie derivative of the action on the jet bundle $\mathscr{J}_k$ is $pr^{(k)}\alpha$ and $d\tilde{x} = dx$ at $\varepsilon = 0$. On the other hand,

$$L_\alpha \, d\tilde{x} = L_\alpha \frac{\partial(\tilde{x}_1, \ldots, \tilde{x}_n)}{\partial(x_1, \ldots, x_n)} \, dx_1, \ldots, dx_n$$

$$= \frac{\delta \partial(X^1, \ldots, X^n)}{\partial(x_1, \ldots, x_n)} \, dx, \ldots, dx_n$$

$$= \left(\sum_{j=1}^n \delta \frac{\partial X^j}{\partial x_j} \right) dx_1, \ldots, dx_n$$

where, as usual, $\delta = d/d\varepsilon|_{\varepsilon=0}$. The last step follows from the fact that $\delta \det A(\varepsilon) = \delta \operatorname{Tr} A(\varepsilon)$ when $A(0) = I$. ($\det A = e^{\operatorname{Tr} \log A}$ for $\| A - I \| < 1$.) Now recall that $\partial X^j/\partial x_j = D_j X^j$ on sections. Therefore

$$\delta \, d\tilde{x} = \sum_{j=1}^n \delta D_j X^j \, dx_1, \ldots, dx_n$$

$$= \sum_{j=1}^n D_j \delta X^j \, dx_1, \ldots, dx_n$$

$$= \sum_{j=1}^n (D_j \xi^j) \, dx_1, \ldots, dx_n.$$

We have found

Theorem 6.7. *The functional* (6.14) *is invariant under the transformation group generated by the vector field α if*

$$(pr^{(k)}\alpha + D_j \xi^j)\omega = 0.$$

Example 6.8. We find the symmetry group of the Lagrangian $(\nabla u)^2\,dx\,dy$. Let $\alpha = \xi(\partial/\partial x) + \eta(\partial/\partial y) + \varphi(\partial/\partial u)$. As before we find a system of differential equations for ξ, η and φ. It is

$$\varphi_x = \varphi_y = 0$$
$$2\varphi_u + \eta_y - \xi_x = 0 \qquad \eta_x + \xi_y = 0$$
$$2\varphi_u + \xi_x - \eta_y = 0 \qquad \xi_u = \eta_u = 0.$$

One finds without difficulty that ξ and η are functions only of x and y and that they satisfy the Cauchy–Riemann equations

$$\xi_x - \eta_y = 0 \qquad \eta_x + \xi_y = 0,$$

while φ is a constant.

Example 6.9. Symmetries of $(\nabla u)^2\,dx_1 \ldots dx_n$ in $\mathbb{R}^n$. The generators and their corresponding group actions are

$$\frac{\partial}{\partial x_i} \qquad\qquad \text{translations}$$

$$x_i \frac{\partial}{\partial x_j} - x_j \frac{\partial}{\partial x_i} \qquad\qquad \text{rotations}$$

$$\frac{\partial}{\partial u} \qquad\qquad u \rightarrow u + c$$

$$-x_i \frac{\partial}{\partial x_i} + \left(\frac{n}{2} - 1\right) u \frac{\partial}{\partial u} \qquad \text{dilations.}$$

Let us compute the group action generated by the last vector field. To get the flow on $X \times U$ we must integrate the differential equations

$$\frac{dx_i}{ds} = x_i \qquad \frac{du}{ds} = \left(\frac{n}{2} - 1\right) u.$$

This leads to the transformation group

$$\tilde{x}_i(s) = e^{-s} x_i \qquad \tilde{u}(s) = e^{((n/2)-1)s} u$$

or, replacing s by $\lambda = e^{-s}$,

$$\tilde{x}_i = \lambda x_i \qquad \tilde{u} = \lambda^{1 - n/2} u.$$

Invariant Solutions

Let $\mathfrak{G}$ be the symmetry group of a differential equation $\Delta = 0$ let, $\mathfrak{H}$ be a subgroup, and suppose the solution $u(x)$ of $\Delta = 0$ is invariant under $\mathfrak{H}$. Then we say that u is an $\mathfrak{H}$-invariant solution. For example, the vector field $\alpha_1 =$

$x(\partial/\partial x) + 2t(\partial/\partial t)$ is an infinitesimal symmetry of the heat equation $u_t - u_{xx} = 0$. The action generated by α, is

$$\tilde{x} = \varepsilon x \qquad \tilde{t} = \varepsilon^2 t \qquad \tilde{u} = u.$$

Therefore $u(x, t)$ is transformed into

$$\tilde{u}(x, t) = u\left(\frac{\tilde{x}}{\varepsilon}, \frac{\tilde{t}}{\varepsilon^2}\right).$$

We may write this transformation group as $(T_\varepsilon u)(x, t) = u(x/\varepsilon, t/\varepsilon^2)$. A function is invariant under this group action if $T_\varepsilon u = u$, that is, if $u(x/\varepsilon, t/\varepsilon^2) = u(x, t)$. Putting $\varepsilon = \sqrt{t}$ we get

$$u\left(\frac{x}{\sqrt{t}}, 1\right) = u(x, t).$$

Let $\xi = x/\sqrt{t}$ and $\varphi(\xi) = u(\xi, 1)$. Then we see that a group invariant solution is $u(x, t) = \varphi(\xi)$. Substituting this form into the last equation we get the ordinary differential equation

$$\varphi'' + \frac{\xi}{2}\varphi' = 0.$$

Two solutions of this equation are $\varphi = \text{const.}$ and $\varphi' = e^{-\xi^2/4}$. Evidently $u_x = (d/dx)\varphi(\xi) = \varphi'(\xi)\,d\xi/dx = (1/\sqrt{t})e^{-x^2/4t}$ is also a solution of the heat equation, and this is the fundamental solution of the heat equation.

For further reading, see Blumen and Cole, Olver, and Ovsjannikov.

EXERCISES

1. Show that the infinitesimal symmetries of the heat equation $u_t - u_{xx} = 0$ are

$$\frac{\partial}{\partial x}, \quad \frac{\partial}{\partial t}, \quad u\frac{\partial}{\partial u}, \quad x\frac{\partial}{\partial x} + 2t\frac{\partial}{\partial t}, \quad 2t\frac{\partial}{\partial x} - xu\frac{\partial}{\partial u},$$

$$4tx\frac{\partial}{\partial x} + 4t^2\frac{\partial}{\partial t} - (x^2 + 2t)u\frac{\partial}{\partial u},$$

$$B(x, t)\frac{\partial}{\partial u}$$

where $B_t - B_{xx} = 0$.

2. Show that the infinitesimal symmetries of the nonlinear system

$$v_y - v_x = 0 \qquad v_y + uu_x = 0$$

are

$$\alpha_1 = (yu^2 - xv)\frac{\partial}{\partial x} - (xu + 2yv)\frac{\partial}{\partial y} + 2uv\frac{\partial}{\partial u} + \left(-\frac{2}{3}u^3 + \frac{3}{2}v^2\right)\frac{\partial}{\partial v}$$

$$\alpha_2 = x\frac{\partial}{\partial x} + 2u\frac{\partial}{\partial u} + 3v\frac{\partial}{\partial v}$$

$$\alpha_3 = \frac{\partial}{\partial v} \qquad \alpha_4 = x\frac{\partial}{\partial x} + y\frac{\partial}{\partial y}.$$

3. Let $\xi(x, y)$ and $\eta(x, y)$ satisfy the Cauchy–Riemann equations. Show that $\xi(\partial/\partial x) + \eta(\partial/\partial y)$ generates a conformal transformation in the plane.

4. Find the infinitesimal symmetries of the nonlinear wave equation $u_{tt} - u_{xx} + u^m = 0$. Show $\alpha = x(\partial/\partial x) + t(\partial/\partial t) + \gamma u(\partial/\partial u)$ is one of them and find the corresponding similarity solutions. In fact, show these take the form $u = t^\gamma \psi(\xi)$ where $\xi = x/t$ and

$$(\xi^2 - 1)\psi'' - 2\gamma\xi\psi' + \gamma(\gamma - 1)\psi + \psi^m = 0.$$

5. Show $[D_i, D_j] = 0$.

6. Find all $F(x, u')$ such that $\alpha = x(\partial/\partial x) + ku(\partial/\partial u)$ is an infinitesimal symmetry of the equation $u'' = F(x, u')$. (Ans. $F = x^{k-2}\varphi(u'/(k-1))$ for arbitrary smooth φ.)

7. Find all g such that differential equation $y' = g(x, y)$ is invariant under the rotation group (6.3). (Ans: $g = (y + x\sigma(r))/(x - y\sigma(r))$ where $\sigma(r)$ is an arbitrary function of $r = \sqrt{x^2 + y^2}$ (see Ince, p. 110).)

CHAPTER 7

Invariant Forms on Lie Groups

24. Invariant Forms on Linear Groups

In the representation theory of finite groups it often is necessary to perform summations over the elements of the group. (See Miller, Chapter 3 or Wigner [2], Chapter 9.) For example, the orthogonality relations for the representations and their characters are expressed as sums over the group. In order to generalize the theory to continuous groups these sums must be replaced by integration with respect to an invariant measure defined on the group. A measure ω is said to be *left invariant* if

$$\int_{\mathfrak{G}} f(gh)\omega = \int_{\mathfrak{G}} f(h)\omega \tag{7.1}$$

for every integrable function f on the group; and the integral (7.1) is called the *Hurwitz* integral. For example, the group of pure rotations can be parametrized by the angle of rotation θ and the axis of rotation $\hat{n}$. All rotations by an angle θ are conjugate, however, and the measure depends only on θ. The invariant integral of a class function χ (one which is constant on conjugacy classes) is given by

$$\frac{1}{\pi} \int_0^\pi \chi(\theta)(1 - \cos \theta) \, d\theta. \tag{7.2}$$

The integral (7.2) is the Hurwitz integral for $SO(3)$.

In this section we develop methods for computing invariant measures on a linear Lie group in local coordinates, as well as invariant forms in general. In §25 we derive the important Maurer–Cartan equations for left invariant forms on a group. These equations are fundamental in the construction of a general group from its algebra (the subject of Chapter 8); but they have other applications as well. Cartan showed how the classical equations of surface theory in $\mathbb{R}^3$ could be derived from the Maurer–Cartan equations for the group of rigid motions; and we shall carry out tnis derivation in §26.

We begin by showing how to calculate left invariant forms in local coordinates. Let L_g denote the diffeomorphism on $\mathfrak{G}$ of left translation by the element g: $L_g h = gh$. As we explained in §19, L_g induces an action on the p-forms of $\mathfrak{G}$ which we denote by ρ_g.

We say that ω is *left invariant* if $\rho_g \omega = \omega$ for all g. For a one-form this means (cf. (5.9)) that $a_i(x)\, dx^i = a_i(x')\, dx'^i$, where $x' = \varphi(x)$ is the coordinate transformation induced by a left translation.

As a simple example, consider the linear group

$$\mathfrak{G} = \left\{ \begin{pmatrix} x & y \\ 0 & 1 \end{pmatrix} \middle| \, x, y \in \mathbb{R} \right\}. \tag{7.3}$$

The transformation of coordinates under left multiplication by $\begin{pmatrix} r & s \\ 0 & 1 \end{pmatrix}$ is given by $x' = rx$, $y' = ry + s$. The one-forms $\omega^1 = x^{-1}\, dx$ and $\omega^2 = x^{-1}\, dy$ are both left invariant. For example

$$\frac{dx'}{x'} = \frac{d(rx)}{rx} = \frac{r\,dx}{rx} = \frac{dx}{x}.$$

Similarly ω^2 is left invariant.

Now if ω^1 and ω^2 are left invariant then so are $d\omega^1$, $d\omega^2$ and $\omega^1 \wedge \omega^2$ by Theorem 5.4. Therefore, in the example above, $x^{-2}\, dx \wedge dy$ is a left invariant measure on the group $\mathfrak{G}$. We leave it to the reader to show that $x^{-1}\, dx \wedge dy$ is a right invariant measure.

We now give an elegant and simple method for constructing a complete set of left invariant one-forms on a linear group $\mathfrak{G}$. Let $x^1, \ldots, x^n$ be a smooth system of local coordinates on a domain $U \subset \mathfrak{G}$ and $B(x^1, \ldots, x^n)$ the corresponding matrix element. Then

Theorem 7.1. *The quantity*

$$\Omega = B^{-1} \frac{\partial B}{\partial x^i} dx^i$$

is a matrix of n linearly independent left invariant one-forms on U.

In the example (7.3) we obtain immediately

$$B^{-1}\, dB = \frac{1}{x}\begin{pmatrix} 1 & -y \\ 0 & x \end{pmatrix}\begin{pmatrix} dx & dy \\ 0 & 0 \end{pmatrix} = \begin{pmatrix} \dfrac{dx}{x} & \dfrac{dy}{x} \\ 0 & 0 \end{pmatrix}.$$

Thus $x^{-1}\, dx$ and $x^{-1}\, dy$ are the left invariant one forms on $\mathfrak{G}$. Similarly a right invariant matrix is given by $(dB)B^{-1}$. In the above case,

$$(dB)B^{-1} = \begin{pmatrix} \dfrac{dx}{x} & \dfrac{-y\,dx + x\,dy}{x} \\ 0 & 0 \end{pmatrix}.$$

PROOF. Let A be a fixed matrix and let $y^i = y^i(x^1, \ldots, x^n)$ be the coordinate transformation induced by left multiplication by A: Thus $B(y^1, \ldots, y^n) = AB(x^1, \ldots, x^n)$. We have

$$B^{-1}(y)\frac{\partial B}{\partial y^j}\,dy^j = B^{-1}(x)A^{-1}\frac{\partial}{\partial y^j}AB(x^1, \ldots, x^n)\,dy^j$$

$$= B^{-1}\frac{\partial B}{\partial x^k}\frac{\partial x^k}{\partial y^j}\,dy^j$$

$$= B^{-1}\frac{\partial B}{\partial x^j}\,dx^j.$$

The independence of the forms reduces easily to the linear independence of the matrices $\partial B/\partial x^1, \ldots, \partial B/\partial x^n$; but these are all tangent vectors to the group at the point B and are independent on U provided the parametrization is regular there. □

Once a set of left-invariant one-forms has been obtained a left invariant measure is given by their wedge product. A left invariant measure could also be obtained directly from the condition

$$a(y)\,dy^1 \wedge \cdots \wedge dy^n = a(x)\,dx^1 \wedge \cdots \wedge dx^n.$$

Substituting $y = y(x)$ on the left side we obtain

$$a(y(x))\frac{\partial(y^1, \ldots, y^n)}{\partial(x^1, \ldots, x^n)} = a(x),$$

where $\partial(y^1, \ldots, y^n)/\partial(x^1, \ldots, x^n)$ is the Jacobian of the transformation $y = y(x)$. In particular, if we choose the coordinate system so that $x = e$ are the coordinates of the identity and normalize the measure so that $a(e) = 1$, we find

$$a(y) = \left.\left(\frac{\partial(y^1, \ldots, y^n)}{\partial(x^1, \ldots, x^n)}\right)^{-1}\right|_{x=e} = \left.\frac{\partial(x^1, \ldots, x^n)}{\partial(y^1, \ldots, y^n)}\right|_{x=e}.$$

For the example (7.3) we found the coordinate transformation to be given by $x' = rx$, $y' = ry + s$. The Jacobian of this transformation is $\partial(x', y')/\partial(x, y) = r^2$, hence at the point x', y' we have $a(x', y') = \partial(x, y)/\partial(x', y')|_{(x=1, y=0)} = 1/r^2 = 1/(x')^2$. The left invariant measure is thus $x^{-2}\,dx \wedge dy$.

Along with left invariant forms we may also construct, in the same way, right invariant forms. As we noted above, $(dB)B^{-1}$ is a matrix of right invariant forms. It follows that every Lie group has both a left and right invariant measure, each unique up to a constant factor. In general, the left and right invariant measures may be different, as in the case of the group (7.3) above; but we may prove

Theorem 7.2. *A compact connected real Lie group has a bi-invariant measure, unique up to a constant factor.*

PROOF. Let R_g be right multiplication on the group: $R_g h = hg$. Then R_g induces a right action τ_g on forms ($\tau_{g_1 g_2} \omega = \tau_{g_2} \tau_{g_1} \omega$). Since the left invariant measure is unique up to a scalar factor we must have $\tau_g \omega = \chi(g) \omega$ where χ is a function on the group. Now $\chi(g_1 g_2) \omega = \tau_{g_1 g_2} \omega = \tau_{g_2} \tau_{g_1} \omega = \chi(g_2) \chi(g_1) \omega$, so $\chi(g_1 g_2) = \chi(g_1) \chi(g_2)$. Moreover, χ never vanishes, since $\chi(\mathbb{1}) = 1$, so χ is a non-trivial homomorphism into the group $\{\hat{R}, \cdot\}$, consisting of the non-zero real numbers with multiplication. Since χ is continuous, its image is a compact connected subgroup of $\{\hat{R}, \cdot\}$; but the only such subgroup of $\{\hat{R}, \cdot\}$ is the group $\{1\}$, so $\chi \equiv 1$. $\qquad\square$

A group need not be compact in order to have a bi-invariant measure. For example, the group of rigid motions in the plane has a bi-invariant measure. (See exercise 1.) A Lie group with a bi-invariant measure is said to be *unimodular*.

We now compute the invariant measure for $SU(2)$ given in the introduction. (See Wigner [2] for an alternative computation.) We parametrize the group $SU(2)$ by the angle of rotation θ and the axis of rotation $\hat{n}$, given by

$$\hat{n} = (\sin \beta \cos \alpha, \sin \beta \sin \alpha, \cos \beta),$$

$$0 \leq \alpha \leq 2\pi, \qquad 0 < \beta < \pi.$$

Recall the spinor representation from Chapter 1 given by $A = e^{-i(\theta/2)\hat{n}\cdot\sigma}$. We find $A = (\cos \theta/2)\mathbb{1} - i\hat{n}\cdot\vec{\sigma} \sin \theta/2$, and

$$A^{-1} dA = \frac{-i}{2} \hat{n}\cdot\vec{\sigma}\, d\theta$$

$$\frac{+i}{\sin \beta} \frac{\partial \hat{n}}{\partial \alpha} \cdot \vec{\sigma} \left(-\sin \frac{\theta}{2} \cos \frac{\theta}{2} \sin \beta\, d\alpha + \sin^2 \frac{\theta}{2} d\beta \right)$$

$$-i \frac{\partial \hat{n}}{\partial \beta} \cdot \vec{\sigma} \left(\sin \frac{\theta}{2} \cos \frac{\theta}{2} d\beta + \sin^2 \frac{\theta}{2} \sin \beta\, d\alpha \right).$$

(Recall that $\hat{n}\cdot\vec{\sigma} = n_1 \sigma_1 + n_2 \sigma_2 + n_3 \sigma_3$.) Since the matrices $i\sigma_j$ all belong to the Lie algebra $su(2)$ we see that $A^{-1} dA$ belongs to $su(2)$. We may therefore write

$$A^{-1} dA = \sum_{j=1}^{3} \omega^j \left(\frac{1}{2i} \sigma_j \right),$$

where ω^j are the left-invariant one-forms

$$\omega^1 = -\sin\beta\cos\alpha\,d\theta - \cos\beta\cos\alpha\sin\frac{\theta}{2}d\beta + \sin\frac{\theta}{2}\sin\alpha\sin\beta\,d\alpha$$

$$\omega^2 = -\sin\alpha\sin\beta\,d\theta - \sin\alpha\sin\frac{\theta}{2}\cos\beta\,d\beta - \cos\alpha\sin\beta\sin\frac{\theta}{2}d\alpha$$

$$\omega^3 = -\cos\beta\,d\theta + \sin\frac{\theta}{2}\sin\beta\,d\beta.$$

The invariant measure is

$$\omega^1 \wedge \omega^2 \wedge \omega^3 = 4\sin\beta\sin^2\frac{\theta}{2}d\alpha \wedge d\beta \wedge d\theta.$$

If we are integrating a class function (one which depends only on θ) the α and β integrations may be carried out, and we arrive at the measure $8\pi\sin^2\frac{\theta}{2} = 4\pi(1 - \cos\theta)\,d\theta$. The normalized measure is

$$\frac{1}{\pi}(1 - \cos\theta)\,d\theta.$$

The Molien function for a representation Γ of a compact group is given by

$$M_\Gamma(z) = \int_{\mathfrak{G}} \det(\mathbb{1} - z\Gamma(g))^{-1}\omega$$

where ω is the normalized invariant measure of the group. Since $\det(\mathbb{1} - z\Gamma(g))^{-1}$ is a class function it suffices to integrate over the conjugacy classes. The Molien function is useful in invariant theory. It is a generating function which counts the number of invariant symmetric tensors of the representation Γ. For applications to problems in physics and bifurcation theory, see Jaric and Birman, and Sattinger [1]. (See also §46.)

EXERCISES

1. The Euclidean group $\mathscr{E}(2)$ can be represented as the matrix group

$$A(x, y, \theta) = \begin{pmatrix} 1 & 0 & 0 \\ x & \cos\theta & \sin\theta \\ y & -\sin\theta & \cos\theta \end{pmatrix}.$$

Find the left and right invariant measures.

2. Find the left and right invariant measures for the solvable group

$$\left\{ \begin{pmatrix} x & z \\ 0 & y \end{pmatrix} \middle| x, y > 0 \right\}.$$

3. Find the left and right invariant measures for the Heisenberg group:

$$\begin{pmatrix} 1 & \alpha & \beta \\ 0 & 1 & \gamma \\ 0 & 0 & 1 \end{pmatrix}.$$

4. The representations of $SU(2)$ are labelled by $j, j = 0, \frac{1}{2}, 1, \ldots$ (see Chapter 4). Show that the Molien function $M_j(z)$ is given by

$$M_j(z) = \frac{1}{\pi} \int_0^\pi \prod_{m=-j}^{j} (1 - ze^{im\theta})^{-1}(1 - \cos\theta)\, d\theta.$$

(*Hint*: Since $\det(\mathbb{1} - z\Gamma(g))$ depends only on the angle of rotation it suffices to compute it for a rotation about the z-axis.) Show

$$M_2(z) = (1 - z^2)^{-1}(1 - z^3)^{-1}$$

$$M_3(z) = \frac{1 + z^{15}}{(1 - z^2)(1 - z^4)(1 - z^6)(1 - z^{10})}.$$

25. The Maurer–Cartan Equations

In the last section we introduced the matrix of left invariant one-forms $\Omega = B^{-1}\, dB$. Let us write this in the form $dB = B\Omega$. Since $d^2 = 0$,

$$d(dB) = 0 = dB \wedge \Omega + B\, d\Omega$$

$$= B(\Omega \wedge \Omega + d\Omega).$$

The expression $\Omega \wedge \Omega$ is to be computed in the obvious way: if $\Omega = (\omega_{ij})$ then $\Omega \wedge \Omega = (v_{ij})$ where $v_{ij} = \sum_k \omega_{ik} \wedge \omega_{kj}$. Since B is non-singular, we have

$$d\Omega + \Omega \wedge \Omega = 0.$$

These are the Maurer–Cartan equations for the group. They can be related to the structure constants of the algebra as follows. We first show that each of the matrices $B^{-1}(\partial B/\partial x^i)$ belongs to the Lie algebra $\mathfrak{g}$. Let $B(t)$ be a curve in the group and let $B_0 = B(0)$, with $B_0 \neq \mathbb{1}$. Then $B_0^{-1}B(t)$ is a curve which passes through the identity at $t = 0$. Its derivative therefore belongs to the Lie algebra, so $B_0^{-1}\dot{B}(0)$ belongs to $\mathfrak{g}$. In particular, the matrices $B^{-1}(\partial B/\partial x^i)$ belong to $\mathfrak{g}$; and, moreover, they form a basis for $\mathfrak{g}$ provided the coordinates $x^1, \ldots, x^n$ are regular—that is, provided the tangent vectors $\partial B/\partial x^i$ are linearly independent in the domain of definition of the variables $x^1, \ldots, x^n$.

The matrix Ω can therefore be written in the form

$$\Omega = \sum_{l=1}^{n} X_l \omega^l$$

where the ω^l are left invariant one-forms on the group and the matrices X_l form a basis for the Lie algebra $\mathfrak{G}$. It is clear that $d\Omega = \sum_{l=1}^{n} X_l\, d\omega^l$. Let us compute $\Omega \wedge \Omega$:

$$\Omega \wedge \Omega = \sum_k X_k \omega^k \wedge \sum_m X_m \omega^m$$

$$= \sum_{k,m} X_k X_m \omega^k \wedge \omega^m$$

$$= \frac{1}{2} \sum_{k,m} [X_k, X_m] \omega^k \wedge \omega^m$$

$$= \frac{1}{2} \sum_{k,m} C^l_{km} X_l \omega^k \wedge \omega^m.$$

where C^l_{km} are the structure constants of the algebra. We may therefore write the Maurer–Cartan equations in the form

$$d\omega^l = -\frac{1}{2} \sum_{k,m} C^l_{km} \omega^k \wedge \omega^m.$$

These are the Maurer–Cartan equations of the group relative to the basis $X_1, \ldots, X_n$.

For example, for the group (7.3) we have $B(x, y) = \left(\begin{smallmatrix} x & y \\ 0 & 1 \end{smallmatrix}\right)$,

$$\Omega = B^{-1} dB = \begin{vmatrix} \dfrac{dx}{x} & \dfrac{dy}{x} \\ 0 & 0 \end{vmatrix} = \omega^1 X_1 + \omega^2 X_2$$

where

$$\omega^1 = \frac{dx}{x}, \qquad \omega^2 = \frac{dy}{x}, \qquad X_1 = \begin{pmatrix} 1 & 0 \\ 0 & 0 \end{pmatrix}, \qquad X_2 = \begin{pmatrix} 0 & 1 \\ 0 & 0 \end{pmatrix}.$$

The Maurer–Cartan equations are

$$d\Omega + \Omega \wedge \Omega = \begin{pmatrix} d\omega_1 & d\omega_2 \\ 0 & 0 \end{pmatrix} + \begin{pmatrix} 0 & x^{-2} dx \wedge dy \\ 0 & 0 \end{pmatrix} = 0,$$

viz.

$$d\omega^1 = 0 \qquad d\omega^2 = -\omega^1 \wedge \omega^2.$$

The preceeding discussion applies to linear groups; but the Maurer–Cartan equations hold for any Lie group. A basis of left invariant forms on a Lie group is constructed as follows. The left action of $\mathfrak{G}$ on itself, defined by $L_g h = gh$, induces an action ρ_g on the vector fields and one-forms of $\mathfrak{G}$. A vector field X on $\mathfrak{G}$ is left invariant if $\rho_g X = X$. Recall that ρ_g carries a tangent vector at p to one at gp. Hence the condition $\rho_g X = X$ means, more precisely, that $\rho_g X_p = X_{gp}$. Similarly, ω is a left invariant one-form if $\rho_g \omega = \omega$, i.e. $\rho_g \omega_p = \omega_{gp}$.

If X is a vector in $\mathfrak{g}$, the Lie algebra of $\mathfrak{G}$, we obtain a left invariant vector field $\tilde{X}$ by setting $\tilde{X}_p = \rho_p X$. Similarly, if ω is a linear functional on $\mathfrak{G}$, $\tilde{\omega}_p = \rho_p \omega$ is a left invariant one-form. Since $(\rho_g \omega)(\rho_g X) = \rho_g \omega(X)$ we see that the function $\tilde{\omega}(\tilde{X})$ is left invariant on $\mathfrak{G}$. But the left action is transitive (it carries every point to every other point); so $\tilde{\omega}(\tilde{X})$ must be constant.

Conversely, if $\tilde{X}$ is a left invariant vector field then $\tilde{X}_p = \rho_p X$, where X is the value of $\tilde{X}$ at the identity; so every left invariant vector field arises in this way. Similarly, every left invariant one-form arises as the left translation of a one-form at the identity. If $X_1, \ldots, X_n$ is a basis for $\mathfrak{g}$ and $\omega^1, \ldots, \omega^n$ a dual basis, then $\tilde{X}_i$ and $\tilde{\omega}_i$ are dual bases of left invariant vector fields and one-forms on $\mathfrak{G}$; and we have $\tilde{\omega}^i(\tilde{X}_j) = \delta_j^i$ everywhere on the group.

If $\tilde{X}$ and $\tilde{Y}$ are two left invariant vector fields given by $\tilde{X}_p = \rho_p X$ and $\tilde{Y}_p = \rho_p Y$, then $[\tilde{X}_p, \tilde{Y}_p] = [\rho_p X, \rho_p Y] = \rho_p[X, Y] = \widetilde{[X, Y]}_p$. This calculation shows that the left invariant vector fields on $\mathfrak{G}$ themselves form a Lie algebra which is isomorphic to $\mathfrak{g}$. If $X_1, \ldots, X_n$ is a basis for $\mathfrak{g}$ and $\tilde{X}_1, \ldots, \tilde{X}_n$ are their left invariant extensions, then

$$[\tilde{X}_j, \tilde{X}_k] = \widetilde{[X_j, X_k]} = \sum_{l=1}^n C_{jk}^l \tilde{X}_l.$$

As in the linear case, we know from Theorem 5.4 that $\tilde{\omega}^i \wedge \tilde{\omega}^j$ and $d\tilde{\omega}^i$ are left invariant. It is clear that the two forms $\{\tilde{\omega}^i \wedge \tilde{\omega}^j | 1 \le i < j \le n\}$ span the space of left invariant two forms, so we have

$$d\tilde{\omega}^i = \sum_{j,k} \gamma_{jk}^i \tilde{\omega}^j \wedge \tilde{\omega}^k$$

for some constants γ_{jk}^i. We may assume that $\gamma_{jk}^i = -\gamma_{kj}^i$ since $\tilde{\omega}^j \wedge \tilde{\omega}^k = -\tilde{\omega}^k \wedge \tilde{\omega}^j$. We leave it as an exercise (see exercise 1 below) to show that $\gamma_{jk}^i = -\frac{1}{2}C_{jk}^i$, where C_{jk}^i are the structure constants for the basis $X_1, \ldots, X_n$.

EXERCISES

1. For any one form θ and vector fields X, Y, show that

$$d\theta(X, Y) = \frac{1}{2}\{X\theta(Y) - Y\theta(X) - \theta([X, Y])\}.$$

   Use this to prove that $\gamma_{jk}^i = -\frac{1}{2}C_{jk}^i$.

2. An n-dimensional manifold is orientable if it supports a nowhere vanishing n-form. Show that every Lie group is an orientable manifold.

3. Let $\mathfrak{G}$ be a Lie group and define $T_g h = ghg^{-1}$. Show that $g \to T_g$ is a homomorphism from $\mathfrak{G}$ into the automorphisms of $\mathfrak{G}$. Show that the action T_g induces an action on the Lie algebra $\mathfrak{g}$ (called the adjoint action) and on the dual space $\mathfrak{g}'$ (called the coadjoint action).

26. Geometry "à la Cartan"

Recall from Chapter 5 that a group action on a manifold naturally induces an action on the vector fields of M, and therefore on its frame bundle. (A frame is a point X on the manifold together with a basis $E_1, \ldots, E_n$ of the tangent space at M; the frame bundle is the manifold of all frames of M.) In this section we discuss a very special class of Riemannian manifolds: those for which the isometry group acts transitively on the frame bundle; that is, every frame of M is carried to every other frame by an isometry. (Here we are considering only the set of oriented, orthonormal frames.) In that case we can identify the frame bundle with the group of isometries. These identifications hold, for example, for R^3, the unit sphere, and the Poincaré half plane.

A manifold M is said to be a *homogeneous space* for a group G if G acts transitively on M, that is, if every point of M is taken to every other point by some transformation in G. Accordingly, we may say that we are discussing Riemannian manifolds M whose frame bundles are homogeneous spaces for the isometry group of M.

All Riemannian spaces are described analytically by a system of partial differential equations, called the structure equations. In the case of the homogeneous spaces described above, these structure equations may be derived from the Maurer–Cartan equations of the isometry group. We shall describe that procedure here. Another example is the derivation of the classical equations of surface theory from the Maurer–Cartan equations of the group of rigid motions of R^3. First proposed by Cartan in 1937 [3], this derivation is outlined in exercise 2 at the end of this section.

A Riemannian manifold is a manifold with a nonsingular, bilinear, symmetric form defined everywhere on its tangent bundle; that is, a bilinear form $g(X, Y)$ where X and Y are vector fields on M. If g is not positive definite, as, for example, in the case of the Minkowski metric in relativity theory, we say the manifold M is a *pseudo-Riemannian* manifold. We sometimes represent the metric tensor as an infinitesimal quantity; for example the Euclidean metric tensor is represented as

$$ds^2 = dx^2 + dy^2 + dz^2.$$

This corresponds to the familiar dot product of Euclidean geometry. If $X = (X_1, X_2, X_3)$ and $Y = (Y_1, Y_2, Y_3)$ then $dx(X) = X_1$, etc; and $ds^2(X, Y) = X_1 Y_1 + X_2 Y_2 + X_3 Y_3$.

In what follows we first restrict our discussion to Riemannian manifolds in which the metric tensor is constant. Such spaces are "flat". We consider the vector space R^n furnished with $ds^2 = g_{ij} dx^i dx^j$, g_{ij} constant, as its metric tensor, and call this space M. Points in M may be represented by $X = (x^1, \dots x^n)$. We define the exterior derivative d on points X by allowing d to operate componentwise: $dX = (dx^1, \dots dx^n)$. Each vector field may also be represented as an n-tuple of functions, that is, as a mapping from R^n to R^n; and the action of the exterior derivative d on vector fields is again defined by operating componentwise. (We could not get away with this on a general curved Riemannian space; in that case we must replace the exterior derivative by the *Riemannian connexion*. We shall discuss this below.)

Let $E_1, \dots E_n$ be a basis of vector fields which span the tangent space at the point X; and let θ_i be the basis of 1-forms dual to the E_i: $\theta_i(E_j) = \delta_{ij}$. For example, we might take $E_i = \partial/\partial x_i$, and $\theta_i = dx_i$. Then

$$dX = (dx_1, \ldots dx_n) = \theta_j E_j.$$

Here we have identified E_i with the tangent vector $(0, \ldots \overset{i}{1}, \ldots 0)$. Thus $dX(E_j) = \sum_k \theta_k(E_j)E_k = E_j$, and dX is the "identity" mapping on the tangent bundle $T(M)$. This same relationship must hold no matter what set of vector fields E_i and dual 1-forms θ_i we choose; so we always have $dX = \theta_j E_j$. Since $d^2 = 0$ we have $d^2 X = 0 = d(\theta_j E_j) = d\theta_j E_j - \theta_j \wedge dE_j$. We next compute dE_j.

Since E_j is a vector field, dE_j is a vector valued 1-form. That is, dE_j is a linear mapping from vector fields to vector fields. For any vector field Y, $dE_j(Y)$ is a vector field; so we may express it as $dE_j(Y) = \omega_{jk}(Y)E_k$ where the coefficients $\omega_{jk}(Y)$ are linear functionals on vector fields—that is, 1-forms. We now have $dX = \theta_j E_j$ and $dE_j = \omega_{jk}E_k$. Since $d^2 = 0$ we get

$$d^2 X = d\theta_j E_j - \theta_j \wedge dE_j = (d\theta_k - \theta_j \wedge \omega_{jk})E_k = 0,$$

$$d^2 E_j = d\omega_{jk}E_k - \omega_{jk} \wedge dE_k = (d\omega_{jm} - \omega_{jk} \wedge \omega_{km})E_m;$$

hence

$$d\theta_k - \theta_j \wedge \omega_{jk} = 0 \qquad d\omega_{jm} - \omega_{jk} \wedge \omega_{km} = 0. \qquad (7.4)$$

These are the *Cartan structural equations* for the Riemannian manifold M. We remind the reader that we have restricted ourselves here to the case of a flat space. We shall discuss the case of manifolds with nonzero curvature below.

There is one additional equation that is satisfied. Since $g(E_i, E_j) = g_{ij}$ is a constant, we have $dg(E_i, E_j) = 0$, and this leads to the condition

$$\omega_{ki}g_{kj} + g_{ik}\omega_{kj} = 0, \qquad \text{i.e.} \quad \omega^t g + g\omega = 0.$$

For example, in R^3 with the usual Euclidean inner product we have $\langle E_i, E_j \rangle = \delta_{ij}$; hence $\omega_{ij} + \omega_{ji} = 0$.

We next identify the isometry group of M with its frame bundle, and derive the structural equations (7.4) from the Maurer–Cartan equations for the isometry group. Every isometry of M is a combination of a translation $(X \to X + a)$ and a "rotation", where by rotation we here mean any matrix U such that $U^t g U = g$, g being the metric tensor. For then $g(R, S) = g(UR, US)$, where R and S are tangent vectors. We represent a frame on M by an $(n + 1) \times (n + 1)$ matrix of the form

$$A = \|X, E_1, \ldots E_n\| = \begin{pmatrix} 1 & 0 \\ X & U \end{pmatrix}$$

where $E_1, \ldots E_n$ is a basis for the tangent space at the point X, and $U = \|E_1, \ldots E_n\|$. Here, 1 is 1 and 0 is really $(0, \ldots 0)$. (Cf. exercise 1, p. 62.) The matrix A may be identified with an isometry as follows. Each point in the space M is represented by a column vector $\binom{1}{Y}$. Under A this point goes into the vector $\binom{1}{UY+X}$. Similarly, tangent vectors are represented by vectors $\binom{0}{V}$, and these transform under A into $\binom{0}{UV}$.

The Maurer–Cartan equations for the isometry group are now easily found. We have

$$dA = \| dX, dE_1, \ldots dE_n \| = \| \theta_i E_i, \omega_{1j_1} E_{j_1}, \omega_{2j_2} E_{j_2}, \ldots \| = A\Omega$$

where

$$\Omega = \begin{pmatrix} 0 & 0 & 0 & \\ \theta_1 & \omega_{11} & \omega_{21} & \cdots \\ \theta_2 & \omega_{12} & \omega_{22} & \\ \vdots & & & \\ \theta_n & \omega_{1n} & \omega_{2n} & \end{pmatrix}.$$

The Maurer–Cartan equations then follow from $d^2 A = 0$; they are simply $d\Omega + \Omega \wedge \Omega = 0$, and it is a simple exercise to show that these give precisely the first and second Cartan structural equations (7.4).

Now let us discuss the case of Riemannian spaces with curvature. For example, the unit sphere in R^3 is a two dimensional Riemannian manifold with constant Gaussian curvature $+1$, while the Poincaré half plane, or the unit disk, discussed in Chapter 1, are Riemannian manifolds with curvature -1. These are examples of "homogeneous" spaces whose symmetry groups act transitively. The symmetry group of the unit sphere (that is, its group of isometries) is the Lie group $SO(3)$; while that of the half plane is $SL(2, R)$. These manifolds are described analytically by a set of structure equations, analogous to equations (7.4) for the flat spaces, but for which the curvature is no longer zero.

In curved spaces we can no longer use our calculations with the exterior derivative and must pass to the more general notion of a *Riemannian connexion*. A *connexion* ∇ is an operation on vector fields X, Y, Z with the following properties:

$$\nabla_X(Y + Z) = \nabla_X(Y) + \nabla_X(Z) \tag{7.5}$$

$$\nabla_{fX} Y = f \nabla_X Y \qquad f \in C^\infty(M)$$
$$\nabla_X fY = (Xf)Y + f \nabla_X Y, \tag{7.6}$$

$$\nabla_{X+Y}(Z) = \nabla_X(Z) + \nabla_Y(Z). \tag{7.7}$$

A *Riemannian* connexion is a connexion ∇ for which

$$Zg(X, Y) = g(\nabla_Z X, Y) + g(X, \nabla_Z Y), \tag{7.8}$$

$$\nabla_X(Y) - \nabla_Y(X) - [X, Y] = 0. \tag{7.9}$$

where g is the metric tensor. When the metric tensor is constant we may obtain the Riemannian connexion from the exterior derivative (in fact, $\nabla_X E_i = dE_i(X)$; but every Riemannian manifold with a non-singular metric tensor g has a unique Riemannian connexion (see Hicks, p. 71)).

We again denote a basis of vector fields for the tangent bundle by $\{E_i\}$.

Since $\nabla_X E_i$ is a vector field we may express it in terms of the basis vector fields $\{E_j\}$:

$$\nabla_X E_i = \omega_{ij}(X)E_j.$$

From the second equation in (7.6) and (7.7) it follows that the ω_{ij} are one forms on M, called the *connexion forms* for ∇. The reader familiar with classical Riemannian geometry will recall that $\nabla_{E_i}(E_i) = \Gamma_{ki}^j E_j$, where the functions Γ_{kj}^j are the Christoffel symbols. Thus $\omega_{ij}(E_k) = \Gamma_{ki}^j$ and the connexion forms ω_{ij} measure the rate of turning of the vector fields E_i as we move from point to point on the manifold. For a concise discussion of the relationship between the various viewpoints of a Riemannian connexion, see Hicks.

For a general vector field $Y = Y^i E_i$ we have from the first equation in (7.6)

$$\nabla_X Y = \nabla_X Y^i E_i = (X Y^i)E_i + Y^i \nabla_X E_i$$

$$= (d Y^j(X) + Y^i \omega_{ij}(X))E_j.$$

Leaving the vector field X unspecified, we may write ∇Y as a vector valued form given by

$$\nabla Y = E_j \otimes (d Y^j + Y^i \omega_{ij}).$$

For those readers familiar with the notion of covariant differentiation of contravariant quantities, it may be useful to point out that the components Y^i of the vector field Y are contravariant quantities, and that their covariant derivative is

$$Y^i_{,j} = \partial Y^i / \partial x^j + Y^k \Gamma^i_{kj}$$

where Γ^i_{kj} are the Christoffel symbols.

Thus ∇Y is a vector-valued form which can be thought of as a mapping from $D^1(M)$ to itself. We may compare ∇ with the exterior derivative d of a C^∞ function f: df is a linear functional from $D^1(M)$ to $C^\infty(M)$; whereas ∇Y is a linear mapping from $D^1(M)$ to itself. For a vector field X, $df[X]$ is the infinitesimal change of f in the direction X; while $\nabla Y[X] = \nabla_X Y$ is the infinitesimal change of the vector field Y in the direction X. The connexion ∇ may be regarded as a linear mapping from $D^1(M) \otimes \Lambda_0$ (recall Λ_0 is identified with $C^\infty(M)$) to linear forms on M with values in $D^1(M)$; that is $\nabla: T(M) \otimes \Lambda_0 \to T(M) \otimes \Lambda_1$. From (7.6) $\nabla(Y \otimes f) = \nabla Y \otimes f + Y \otimes df$.

There are two tensors associated with a connexion, the torsion and the curvature:

$$T(X, Y) = \nabla_X(Y) - \nabla_Y(X) - [X, Y], \tag{7.10}$$

$$R(X, Y) = \nabla_X \nabla_Y - \nabla_Y \nabla_X - \nabla_{[X, Y]}. \tag{7.11}$$

The condition (7.9) states that a Riemannian connexion has zero *torsion*.

It is clear that $T(X, Y)$ is a vector field and that $R(X, Y)$ is an operation on vector fields. It is a straightforward calculation to show that $T(fX, gY) = fg T(X, Y)$ and that $R(fX, gY)hZ = (fgh)R(X, Y)Z$, where f, g, h are in $C^\infty(M)$.

This shows that T and R depend only on the values of the vector fields at the point on the manifold; for we may multiply these vector fields by C^∞ functions which are identically one in a neighborhood of a fixed point p and identically zero outside a small neighborhood of p, and the values of T and R at p will be unaffected. This proves that T and R are *tensors* on M; that is, that they depend only on the values of the vector fields at the particular point of M.

It is also clear that T and R are antisymmetric in X and Y and so may be represented in terms of differential forms. We define the torsion and curvature forms T_k and R_{kl} for ∇ by

$$T(E_i, E_j) = \sum_k T_k(E_i, E_j)E_k$$

$$R(E_i, E_j)E_k = \sum_{kl} R_{kl}(E_i, E_j)E_l.$$

If the $\{E_i\}$ form an orthonormal basis relative to a metric tensor g, then the dual forms θ_i are given by $\theta_i(X) = g(X, E_i)$. The connexion, torsion, and curvature forms are all related by equations (7.10) and (7.11); the resulting equations are called the first and second structural equations of the connexion ∇:

Theorem 7.3. *The structural equations for a connexion are:*

$$\tfrac{1}{2}T_k = d\theta_k - \theta_j \wedge \omega_{jk} \qquad \text{first structural equations}$$

$$\tfrac{1}{2}R_{mn} = d\omega_{mn} - \omega_{mj} \wedge \omega_{jn} \qquad \text{second structural equations.}$$

PROOF. To derive the first structural equation we need the identity

$$2\,d\theta(X, Y) = X\theta(Y) - Y\theta(X) - \theta([X, Y]).$$

(See exercise 1, §25.) We begin by writing

$$\nabla Y = \sum_k E_k \otimes (dY^k + Y^j\omega_{jk}) = \sum_k E_k \otimes (d\theta_k(Y) + \omega_{jk}\theta_j(Y)).$$

Then

$$\nabla_X Y - \nabla_Y X = \sum_k E_k(dY^k(X) - dX^k(Y) + Y^j\omega_{jk}(X) - X^j\omega_{jk}(Y))$$

$$= \sum_k E_k(X\theta_k(Y) - Y\theta_k(X) - 2\theta_j \wedge \omega_{jk}(X, Y))$$

$$= \sum_k E_k(2\,d\theta_k(X, Y) - 2\theta_j \wedge \omega_{kj}(X, Y) + \theta_k([X, Y]))$$

from our identity. Hence

$$\nabla_X Y - \nabla_Y X - [X, Y] = 2\sum_k E_k(d\theta_k - \theta_j \wedge \omega_{jk})(X, Y);$$

and the first structural equation follows immediately from the definition of the torsion forms T_k.

The second structural equation is proved by showing that

$$R(E_j, E_k)E_i = 2\sum_m E_m(d\omega_{im} - \omega_{in} \wedge \omega_{nm})(E_j, E_k);$$

this is a straightforward calculation and is left as an exercise.   □

Whereas $d^2 = 0$, it is not in general true that $\nabla^2 = 0$ for a connexion; and the curvature tensor is associated with the (in general) nonvanishing of ∇^2. We must first explain how ∇^2 is defined. As we noted above, $\nabla \colon D^1(M) \otimes \Lambda_0 \to D^1(M) \otimes \Lambda_1$. We extend ∇ to a mapping from $D^1(M) \otimes \Lambda_1$ to $D^1(M) \otimes \Lambda_2$ by requiring that $\nabla X \otimes \omega = \nabla X \wedge \omega + X \otimes d\omega$. Then

$$\nabla^2 E_i = \nabla(E_j \otimes \omega_{ij}) = \nabla E_j \wedge \omega_{ij} + E_j \otimes d\omega_{ij}$$

$$= E_k \otimes (\omega_{jk} \wedge \omega_{ij} + d\omega_{ik})$$

$$= E_k \otimes (d\omega_{ik} - \omega_{ij} \wedge \omega_{jk}).$$

Thus $\nabla^2 E_i$ is a vector valued two form; and $(\nabla^2 E_i)(E_j, E_k) = \frac{1}{2}R(E_j, E_k)E_i$.

As a result, $\nabla^2 = 0$ only if the curvature is zero. In this case the connexion is said to be *flat*. The Riemannian connexion in Euclidean space, the Minkowski metric, and in fact any constant metric tensor have flat connexions. These cases were discussed in the first part of the section. Then, the Riemannian connexion is obtained by extending the exterior derivative d to vector fields, and the structural equations (7.4) follow immediately from the relation $d^2 = 0$. We showed above that these equations could be derived from the Maurer–Cartan equations of the isometry group. We now want to carry out a similar derivation for curved "homogeneous" spaces.

We first discuss the Cartan structural equations for a two-dimensional smooth surface S embedded in R^3. It becomes a Riemannian manifold by restricting the Euclidean metric to the tangent bundle $T(S)$. In other words, if X and Y are two tangent vectors to S, we define $g(X, Y) = \langle X, Y \rangle$, where $\langle \; , \; \rangle$ is the ordinary Euclidean inner product in R^3. Let a patch of the surface S be parametrized by u, v; and let $X(u, v)$ be a point on S. At each point X let $E_1(u, v)$ and $E_2(u, v)$ be orthonormal tangent vector fields, and let $E_3 = E_1 \times E_2$ be the normal vector field (see Figure 7.1).

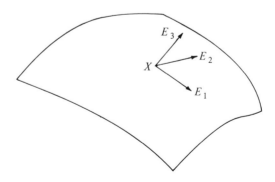

Figure 7.1

Since X, E_1, E_2, and E_3 sit in R^3 we may define the exterior derivatives dX and dE_i in the ambient space R^3. We write $dX = \theta_i E_i$ and $dE_i = \omega_{ij} E_j$. Now $dX = X_u\, du + X_v\, dv$; and since X_u and X_v are tangent to S, $dX \cdot E_3 = 0$. Therefore $\theta_3 = 0$ and $dX = \theta_1 E_1 + \theta_2 E_2$. The Riemannian connexion for the surface S is obtained by taking the component of the Riemannian connexion for R^3 which is tangential to S. For R^3 the Riemannian connexion is $dE_i = \omega_{ij} E_j$. For example $dE_1 = \omega_{12} E_2 + \omega_{13} E_3$; so the Riemannian connexion for S is given by $\nabla E_1 = \omega_{12} E_2$, $\nabla E_2 = \omega_{21} E_1$. (Recall that $\omega_{12} = -\omega_{21}$.) The structural equations for the surface S, considered as a Riemannian manifold, are (see O'Neill, p. 276)

$$d\theta_1 = \omega_{12} \wedge \theta_2 \qquad d\theta_2 = \omega_{21} \wedge \theta_1$$
$$d\omega_{12} = -K\theta_1 \wedge \theta_2, \tag{7.12}$$

where K is its Gaussian curvature.

These equations may be derived from the Maurer–Cartan equations of the isometry group for certain "homogeneous" spaces. For example, let us derive the structural equations for a sphere of radius 1 in R^3. Points in the frame bundle of the unit sphere may be represented by three orthonormal vectors E_1, E_2, and E_3, where E_3 is normal to the sphere and E_1 and E_2 are tangent to the sphere, as below.

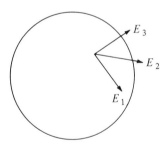

We identify E_3 with a point on the sphere itself (i.e. E_3 plays the role of X in the discussion above). Setting $O = \|E_1, E_2, E_3\|$, where $O \in SO(3)$, we have, since $dE_i = \omega_{ij} E_j$,

$$dO = \|dE_1, dE_2, dE_3\| = \|E_1, E_2, E_3\| \Omega,$$

where

$$\Omega = \begin{pmatrix} 0 & \omega_{21} & \omega_{31} \\ \omega_{12} & 0 & \omega_{32} \\ \omega_{13} & \omega_{23} & 0 \end{pmatrix}.$$

Since $dE_3 = \omega_{31} E_1 + \omega_{32} E_2$, we set $\theta_1 = \omega_{31}$ and $\theta_2 = \omega_{32}$. Since $dE_1 = \omega_{12} E_2 + \omega_{13} E_3$, the Riemannian connexion is given by $\nabla E_1 = \omega_{12} E_2$ and we identify the Maurer–Cartan form ω_{12} and the connexion form ω_{12} as one and the same. With these identifications we write

$$\Omega = \begin{pmatrix} 0 & \omega_{21} & \theta_1 \\ \omega_{12} & 0 & \theta_2 \\ -\theta_1 & -\theta_2 & 0 \end{pmatrix}.$$

The structure equations for the unit sphere then follow from the Maurer–Cartan equations $d\Omega + \Omega \wedge \Omega = 0$. These are equations (7.12) with $K = 1$.

In the exercises below we outline similar derivations along these lines, one for the derivation of the structure equations of a Riemannian manifold of constant negative curvature; and one for the classical equations of surface theory, in the form descended from Cartan [3].

EXERCISES

1. Consider R^3 with the metric tensor

$$g = \begin{pmatrix} 1 & 0 & 0 \\ 0 & 1 & 0 \\ 0 & 0 & -1 \end{pmatrix}.$$

Consider the hyperboloid H given by $g(X, X) = -1$, where $X = (x, y, z)$; i.e. $x^2 + y^2 - z^2 = -1$.

(i) Show the frame bundle for H is the group $SO(2, 1)$
(ii) Let $O \in SO(2, 1)$ and let $\Omega = O^{-1} dO$. Show that $\Omega g + g\Omega^t = 0$.
(iii) Derive the structure equations for the manifold H and show that it has constant Gaussian curvature -1.

2. Let S be a surface embedded in R^3, and suppose it is parametrized locally by the variables u, v. Let $\| X, E_1, E_2, E_3 \|$ be an adapted frame field on S; that is, X and E_i are smooth functions of u and v, and E_1, E_2, E_3 is a right-handed orthonormal frame at X with $E_3 = E_1 \times E_2$ normal to S.

(i) Show that such an adapted frame field can be regarded as a smooth function with values in the group of rigid motions of R^3 by showing that the 4×4 matrix given by

$$A = \begin{pmatrix} 1 & 0 \\ X & U \end{pmatrix},$$

where $U = \| E_1, E_2, E_3 \| \in SO(3)$, is an element of $\mathscr{E}(3)$.
(ii) Show that $dX = \theta_1 E_1 + \theta_2 E_2$, where θ_1 and θ_2 are the dual forms to the tangent vectors E_1 and E_2.
(iii) Let $\Omega = A^{-1} dA$. Show that

$$\Omega = \begin{pmatrix} 0 & 0 & 0 & 0 \\ \theta_1 & 0 & \omega_{21} & \omega_{31} \\ \theta_2 & \omega_{12} & 0 & \omega_{32} \\ 0 & \omega_{13} & \omega_{23} & 0 \end{pmatrix}.$$

(iv) From the Maurer–Cartan equations for $\mathscr{E}(3)$ derive the following structure equations for the embedded surface S:

$$d\theta_1 = \omega_{12} \wedge \theta_2 \qquad \text{first structural equations}$$

$$d\theta_2 = \omega_{21} \wedge \theta_1$$

$$\omega_{31} \wedge \theta_1 + \omega_{32} \wedge \theta_2 = 0 \qquad \text{symmetry equation}$$

$$d\omega_{12} = \omega_{13} \wedge \omega_{32} \qquad \text{Gauss equation}$$

$$d\omega_{13} = \omega_{12} \wedge \omega_{23} \qquad \text{Codazzi equations}$$

$$d\omega_{23} = \omega_{21} \wedge \omega_{13}$$

These are the classical equations of surface theory (see O'Neill, p. 249 for a direct geometric derivation). The philosophy of this approach to their derivation is suggested in Cartan [3], though the equations do not appear in the explicit form given here. See also Flanders.

3. The unit disk becomes a two dimensional Riemannian manifold M with constant negative curvature with the metric $ds^2 = dz\, dz^*/(1 - |z|^2)^2$. Recall from Chapter 1 that $SU(1,1)/\pm 1$ is the isometry group of the unit disk with this metric, where

$$SU(1,1) = \left\{ \begin{pmatrix} \alpha & \beta \\ \beta^* & \alpha^* \end{pmatrix} \;\middle|\; |\alpha|^2 - |\beta|^2 = 1 \right\}.$$

(i) Show that the frame bundle of M can be represented by pairs of complex numbers $\{z, w\}$ where $|z| < 1$, and $|w|^2 = (1 - |z|^2)$.
(ii) Show that the frame bundle can be identified with $SU(1,1)$ via $\alpha = (v^*)^{-1/2}$ and $\beta = z/(v)^{1/2}$.
(iii) Let θ_1, θ_2, and ω_{12} be the forms for the surface M; show how to identify them with the Maurer–Cartan forms for the group $SU(1,1)$, and derive the structural equations for M.

27. Variations on a Theme by Euler

Recall (§14) that the configuration space for the motion of a rigid body is the Lie group $SO(3)$ and that Euler's equations (4.5) are obtained by bringing the equations down to the Lie algebra $so(3)$. V.I. Arnold [1] showed how the Lie group $SO(3)$ could be equipped with a left invariant metric tensor in such a way that the trajectories of the motion are geodesics on the resulting Riemannian space. He then showed how Euler's equations of motion for an inviscid incompressible fluid could be treated, at least formally, from the same point of view. In the case of the fluid the group is the infinite dimensional group of volume preserving diffeomorphisms, and the algebra is the Lie algebra of divergence free vector fields.

In this section we discuss briefly Arnold's ideas as an illustration of the concepts introduced in §24. The reader interested in pursuing the topic further may consult Arnold's original article [1] or his book (Arnold [2], pp. 301–

342). (See also Ebin and Marsden for a hydrodynamical existence theorem based on Arnold's ideas.)

The left invariant metric on the group $SO(3)$ which is compatible with the motion of the rigid body is obtained directly from the kinetic energy, *viz.*

$$T = \tfrac{1}{2}m(I\Omega, \Omega)$$

where Ω is the angular velocity ($\Omega = O^t\dot{O}$, $O(t)$ being the frame of the body at time t), I is the inertia operator, and $(\ ,\ )$ is the Euclidean inner product. From now on we take the mass to be unity. Recall that we identified $\mathbb{R}^3$ with the Lie algebra $so(3)$ by $\Omega \to \Omega^j L_j$. The Euclidean inner product is given in terms of the Killing form by $(X, Y) = -\tfrac{1}{2}K(X, Y)$. Thus we take for the metric tensor on the algebra $so(3)$

$$g(X, Y) = -\tfrac{1}{2}K(IX, Y). \tag{7.13}$$

Since g is a symmetric bilinear form we can choose a basis E_i for $so(3)$ such that $IE_j = I_j E_j$ and $[E_i, E_j] = \varepsilon_{ijk} E_k$.

The left invariant metric on the group is then obtained by left translation of g from the identity: if X and Y are tangent vectors to $SO(3)$ at O then $g(X, Y) = g(O^t X, O^t Y)$.

The equations of the geodesics in a Riemannian space can be obtained from the Riemannian connexion. A geodesic of the connexion ∇ is a curve $\gamma(t)$ whose tangent vector, $\dot{\gamma}$, is parallel along the curve, thus $\nabla_{\dot{\gamma}}\dot{\gamma} = 0$. In local coordinates this condition leads to the ordinary differential equations

$$\ddot{\gamma}^j + \Gamma^j_{ik}\dot{\gamma}^i\dot{\gamma}^k = 0,$$

where the Christoffel symbols Γ^k_{ij} are smooth functions on M such that

$$\nabla_{\partial_i}\partial_j = \Gamma^k_{ij}\partial_k, \qquad \partial_i = \frac{\partial}{\partial x^i}.$$

In the case we are considering, that of left invariant metrics, the connexion will also be left invariant, *viz.* $\nabla_{\tilde{X}}\tilde{Y} = \widetilde{\nabla_X Y}$ (see exercise 1). If $X_1, \ldots, X_n$ is a basis for the Lie algebra $\mathfrak{g}$ and $\nabla_{X_i} X_j = \Gamma^k_{ij} X_k$, then $\nabla_{\tilde{X}_i}\tilde{X}_j = \Gamma^k_{ij}\tilde{X}_k$ for the left invariant vector fields $\tilde{X}_i$.

Let us return now to the metric tensor given by (7.13). Since the metric tensor g is left invariant, $g(\tilde{X}, \tilde{Y})$ is constant for any pair of left invariant vector fields $\tilde{X}$ and $\tilde{Y}$; and (7.8) reduces to the condition

$$K(I\nabla_Z X, Y) + K(IX, \nabla_Z Y) = 0 \tag{7.14}$$

for any matrices X, Y, Z in $so(3)$.

Choose as a basis for $so(3)$ the matrices E_j which are eigenvectors of I and let $\nabla_{E_i} E_j = \Gamma^k_{ij} E_k$. Equations (7.14) give a system of equations for the quantities Γ^k_{ij}. We leave it to the reader to show they are given by

$$\Gamma^k_{ij} = \frac{1}{2}\varepsilon_{ijk}\left(1 - \frac{I_i - I_j}{I_k}\right). \tag{7.15}$$

Now let us show that Euler's equations (4.5) are the equations of the geodesics in $SO(3)$ with the metric we have chosen. The curve in $SO(3)$ is given by a matrix $O(t)$, and $\Omega = O^t \dot{O} \in so(3)$. Writing $\Omega = \Omega^j E_j$ we find $\dot{O} = \Omega^j(t) O E_j = \Omega^j \tilde{E}_j$, the matrices $\tilde{E}_j = O E_j$ being the left invariant vector fields on the group. We have

$$\nabla_{\dot{O}} \dot{O} = \nabla_{\Omega^j \tilde{E}_j} \Omega^k \tilde{E}_k = [(\Omega^j \tilde{E}_j) \Omega^k] \tilde{E}_k + \Omega^j \Omega^k \nabla_{\tilde{E}_j} \tilde{E}_k.$$

Now recall that by $(\Omega^j \tilde{E}_j) \Omega^k$ we mean the tangent vector $\Omega^j \tilde{E}_j$ acting on the function Ω^k as a derivation—that is, as differentiation along the curve. Therefore $\Omega^j \tilde{E}_j \Omega^k = d\Omega^k/dt$, and so

$$\nabla_{\dot{O}} \dot{O} = \left(\frac{d\Omega^l}{dt} + \Gamma^l_{jk} \Omega^j \Omega^k \right) \tilde{E}_l.$$

Thus the equations for the geodesics are

$$\dot{\Omega}^l + \Gamma^l_{jk} \Omega^j \Omega^k = 0.$$

We leave it to the reader to verify that these are precisely (4.5) when Γ^l_{jk} are given by (7.15).

We close this section with a brief account of Arnold's generalization of this point of view to Euler's equations of motion of an inviscid incompressible fluid. Those equations are

$$\frac{\partial u^i}{\partial t} + u^j \frac{\partial u^i}{\partial x^j} = -\frac{\partial p}{\partial x^i}$$

$$\frac{\partial u^i}{\partial x^i} = 0.$$

(7.16)

Here p is the hydrodynamical pressure and u^1, u^2, u^3 are the components of the fluid velocity. The fourth equation expresses the fact that the fluid is incompressible.

The instantaneous configuration of the fluid is specified by giving the positions of the fluid particles relative to their initial positions. Since the fluid is incompressible, this is a volume preserving transformation, so the configuration space of the fluid is the group of volume preserving diffeomorphisms of the region occupied by the fluid. (Let us assume that we are dealing with smooth C^∞ solutions of (7.16).) The transformation is obtained by integrating the differential equations

$$\frac{dx^i}{dt} = u^i(x, t).$$

The flow obtained by solving these equations is volume preserving since u^i is a divergence free vector field. (This is a special case of exercise 2, §20.) We may say then, at least formally, that the Lie algebra of the group of volume preserving diffeomorphisms is the algebra of divergence free vector fields.

As in rigid body motion, Euler's equations for the fluid take place in the algebra. Again, in analogy with the rigid body case, the appropriate metric tensor is the kinetic energy, *viz.*

$$\frac{1}{2}\iint_B u^i u^i \, dx$$

where the integral is carried out over the region B occupied by the fluid. The equations of the flow are again geodesics relative to this metric tensor.

Arnold further calculated the curvature tensor for this metric tensor and used the results to speculate about the stability of fluid flow and the difficulties inherent in meteorological predictions. (See the discussion in Appendix 2 of his book [2].)

EXERCISES

1. Let $\mathfrak{G}$ be a Lie group with affine connexion ∇ and left invariant metric tensor g. Let $\{\tilde{E}_i\}$ be an orthonormal basis of left invariant vector fields on $\mathfrak{G}$ ($g(\tilde{E}_i, \tilde{E}_j) = \delta_{ij}$). Show that the Christoffel symbols are given by

$$\Gamma^k_{ij} = \tfrac{1}{2}(C^k_{ij} + C^j_{ki} - C^i_{jk})$$

where the C^k_{ij} are the structure constants of the algebra. *Hint*: Use (7.8) and (7.9). Hence show that ∇ is left invariant.

2. Show that relative to a basis of normalized eigenvectors the Christoffel symbols in (7.15) become

$$\Gamma^k_{ij} = \left(\frac{I_j + I_k - I_i}{\sqrt{2I_i I_j I_k}}\right)\varepsilon_{ijk}.$$

CHAPTER 8

Lie Groups and Algebras: Differential Geometric Approach

28. The Maurer–Cartan Equations, *bis*

In Chapter 3 we constructed a local Lie group from a linear Lie algebra by exponentiating the matrices in the algebra. By Ado's theorem, this method succeeds in obtaining all local Lie groups; but Lie's original methods involved the integration of overdetermined systems of partial differential equations. The classical solution of the problem is quite involved, though Pontryagin gives a fairly concise treatment of it. The calculations are simplified considerably by Cartan's use of the calculus of differential forms (exterior calculus). We shall develop the subject in this chapter, comparing both the classical and modern approaches, and giving efficient proofs of the basic results using the exterior calculus. This chapter is included purely for its historical interest, and is independent of the rest of the book.

We begin by defining an analytic manifold.

Let M be a topological space. A *local chart* on M is a pair (U, ϕ) where U is an open set in M and ϕ is a homeomorphism of U onto an open set $\phi(U)$ in R^n for some n. The local chart (U, ϕ) may be viewed as a system of local coordinates on M defined on the set U. Each p in U has a set of coordinates $x^i(p) = \phi^i(p), i = 1, \ldots, n$. A *smooth manifold* M is a topological space together with a collection of local charts (ϕ_α, U_α) (called an *atlas*) such that the $\{U_\alpha\}$ cover M and for each pair α, β the mapping $\phi_\beta \circ \phi_\alpha^{-1}$ is a C^∞ mapping from $\phi_\alpha(U_\alpha \cap U_\beta)$ to $\phi_\beta(U_\alpha \cap U_\beta)$. An analytic manifold is a smooth manifold for which the mapping $\phi_\alpha \circ \phi_\beta^{-1}$ is analytic (*i.e.* $\phi_\alpha \circ \phi_\beta^{-1}$ is represented by a convergent Taylor series in a neighborhood of each x in $\phi_\beta(U_\alpha \cap U_\beta)$.)

A mapping $f: M \to N$, where M and N are analytic, is said to be analytic if for every pair of charts (U, ϕ) on M and (V, ψ) on N, $\phi \circ f \circ \psi^{-1}$ is an analytic mapping from $\psi(V)$ to $\phi(U)$. A Lie group $\mathfrak{G}$ is an analytic manifold which is also a group and for which the operation $(A, B) \to A \circ B^{-1}$ from $\mathfrak{G} \times \mathfrak{G}$ to $\mathfrak{G}$ is analytic. A local Lie group is an analytic manifold with the group structure defined only locally in some neighborhood of the identity element.

For any system of local coordinates in the neighborhood of the identity of a Lie group $\mathfrak{G}$, the product in $\mathfrak{G}$ induces a mapping h of $R^n \times R^n$ into R^n as follows. Write $g(x)$ for the element in $\mathfrak{G}$ with coordinates $x \in R^n$, and assume $g(0) = \mathbb{1}$ (the identity in $\mathfrak{G}$). The mapping h is defined by

$$g(h(x, y)) = g(x)g(y)$$

where $g(x)g(y)$ denotes the product of the two elements $g(x)$ and $g(y)$. The mapping h is defined on a sufficiently small neighborhood of the origin in $R^n \times R^n$. By associativity we have

$$h(h(x, y), z) = h(x, h(y, z)). \tag{8.1}$$

Moreover, $h(0, x) = h(x, 0) = x$, since the origin is mapped into $\mathbb{1}$.

Since the product is analytic, h is analytic, and so may be expanded in a Taylor series:

$$h^i(x, y) = x^i + y^i + \sum_{r,s} a^i_{rs} x^r y^s + \cdots. \tag{8.2}$$

Let x^{-1} denote the coordinates of the point $(g(x))^{-1}$. It is defined implicitly by the equation $h(x, x^{-1}) = 0$. Since $\partial h^i / \partial y^j(0,0) = \delta^i_j$ the equation $h(x, y) = 0$ can be solved uniquely for y by the implicit function theorem. From (8.2) we obtain the expansion

$$(x^{-1})^i = -x^i + \sum_{r,s} a^i_{rs} x^r x^s + \cdots. \tag{8.3}$$

We now construct a Lie product on R^n, making it the Lie algebra of the group. Given $x, y \in R^n$ define $q(x, y)$ in $\mathfrak{G}$ by

$$q(x, y) = g(x)g(y)g^{-1}(x)g^{-1}(y)$$

and let $q^i(x, y)$ be the local coordinate of q. We have $q^i(x, y) = h^i(h(x, y), h^{-1}(y, x))$, and using the Taylor expansion (8.2) we get

$$q^i(x, y) = h^i(x, y) + (h^{-1}(y, x))^i$$
$$= x^i + y^i + a^i_{rs} x^r y^s - y^i - x^i - a^i_{rs} y^r x^s$$
$$= c^i_{rs} x^r y^s + \cdots$$

where $c^i_{rs} = a^i_{rs} - a^i_{sr}$.

The c^i_{rs} are the structure constants of the Lie algebra. They are clearly skew-symmetric; and we shall prove below that as a consequence of the associativity (8.1) of the product, they satisfy the Jacobi identity.

We next derive a system of differential equations which h must satisfy. Let

$$A_j^i(x) = \frac{\partial h^i}{\partial y^j}(x^{-1}, y)|_{y=x}. \tag{8.4}$$

For fixed z we put $y = h(z, x)$ and $y_0 = h(z, x_0)$. Then $x \to y = h(z, x)$ corresponds to left multiplication in the group by $g(z)$. By the associativity (8.1) we find that

$$h(y_0^{-1}, y) = h(h^{-1}(z, x_0), h(z, x))$$
$$= h(h(h^{-1}(z, x_0), z), x)$$
$$= h(x_0^{-1}, x).$$

Differentiating both sides with respect to x^l we get

$$\sum_k \frac{\partial h^i}{\partial y^k}(y_0^{-1}, y)\frac{\partial h^k}{\partial x^l}(z, x) = \frac{\partial h^i}{\partial x^l}(x_0^{-1}, x),$$

and putting $x = x_0$ and $y = y_0$ we find

$$\sum_k A_k^i(y_0)\frac{\partial h^k}{\partial x^l}(z, x_0) = A_l^i(x_0).$$

Dropping the subscripts and recalling that $y = h(z, x)$, we put these equations in the form

$$\sum_k A_k^i(h)\frac{\partial h^k}{\partial x^l}(z, x) = A_l^i(x) \tag{8.5}$$

identically in z. From (5.9') we see that (8.5) is a statement of the fact that the one forms

$$\omega^i = \sum_{l=1}^{n} A_l^i(x)\, dx^l$$

are *left invariant*. These forms are known as the *Maurer–Cartan* forms.

The Maurer–Cartan forms are a basis for the left invariant one forms on $\mathfrak{G}$. By our arguments in §25 we have

$$d\omega^i = \sum_{j,k} \gamma_{jk}^i \omega^j \wedge \omega^k$$

where γ_{jk}^i are constants. Since $\omega^j \wedge \omega^k = -\omega^k \wedge \omega^j$ we may assume that $\gamma_{jk}^i + \gamma_{kj}^i = 0$.

Let us show that $\gamma_{jk}^i = -\frac{1}{2}c_{jk}^i$ where this time $c_{rs}^i = a_{rs}^i - a_{sr}^i$. From (8.2) and (8.4)

$$A_j^i(x) = \delta_j^i + \sum_r a_{rj}^i(x^{-1})^r$$
$$= \delta_j^i - \sum_r a_{rj}^i x^r + \cdots.$$

Therefore $\omega^i(x) = dx^i - \sum_{r,j} a^i_{rj} x^r \, dx^j + \cdots$, and

$$d\omega^i = -\sum_{j,k} a^i_{jk} \, dx^j \wedge dx^k + \cdots$$

$$= -\sum_{j<k} (a^i_{jk} - a^i_{kj}) \, dx^j \wedge dx^k + \cdots$$

$$= -\sum_{j<k} c^i_{jk} \, dx^j \wedge dx^k + \cdots.$$

On the other hand, since γ^i_{jk} are skew symmetric in j and k,

$$\sum_{j,k} \gamma^i_{jk} \omega^j \wedge \omega^k = \sum_{j,k} \gamma^i_{jk} \, dx^j \wedge dx^k + \cdots$$

$$= 2 \sum_{j<k} \gamma^i_{jk} \, dx^j \wedge dx^k + \cdots.$$

Comparing these two expressions for $d\omega^i$ we see that $\gamma^i_{jk} = -\tfrac{1}{2} c^i_{jk}$.
We have thus again derived the *Maurer–Cartan* equations

$$d\omega^i = -\frac{1}{2} \sum_{j,k} c^i_{jk} \omega^j \wedge \omega^k. \tag{8.6}$$

We can now prove

Theorem 8.1. *The structure constants c^i_{jk} satisfy the Jacobi identity and so determine a Lie algebra.*

PROOF. Since $d^2 = 0$ we have

$$d(d\omega^i) = 0 = -\frac{1}{2} \sum_{j,k} c^i_{jk} (d\omega^j \wedge \omega^k - \omega^j \wedge d\omega^k)$$

$$= -\sum_{j,k} c^i_{jk} \, d\omega^j \wedge \omega^k$$

$$= \frac{1}{2} \sum_{j,k,r,s} c^i_{jk} c^j_{rs} \omega^r \wedge \omega^s \wedge \omega^k$$

$$= \frac{1}{6} \sum_{k,r,s} \Gamma^i_{krs} \omega^r \wedge \omega^s \wedge \omega^k$$

where

$$\Gamma^i_{krs} = \sum_j c^i_{jk} c^j_{rs} + c^i_{jr} c^j_{sk} + c^i_{js} c^j_{kr}.$$

Here we have used the fact that $\omega^r \wedge \omega^s \wedge \omega^k = \omega^s \wedge \omega^k \wedge \omega^r$, etc. We may write the previous quantity as

$$\sum_{k<r<s} \Gamma^i_{krs} \omega^k \wedge \omega^r \wedge \omega^s$$

since $\Gamma^i_{krs} = \Gamma^i_{rsk} = \Gamma^i_{skr}$. But this implies that $\Gamma^i_{krs} = 0$ since $\{\omega^k \wedge \omega^r \wedge \omega^s | k < r < s\}$ are independent 3-forms. Thus the c^i_{jk} satisfy the Jacobi identity. $\qquad\square$

29. Construction of the Group from the Algebra

We have now derived the Lie algebra from the group. The multiplication on the group induced the composition function $h(x, y)$ in the local coordinates of the group. The structure constants then come from the second derivatives of h. Conversely, given a set of structure constants we are going to construct a Lie group by solving (8.5) for the composition function h. The solvability conditions for (8.5) are the Maurer–Cartan equations, which contain the structure constants. The program to construct the Lie group from the algebra (as represented by the structure constants) is as follows:

(i) Solve the Maurer–Cartan equations given the structure constants and obtain the Maurer–Cartan forms.
(ii) Solve equations (8.5) for the composition function.
(iii) Show h is associative.

We begin by showing that any solution of (8.5) is associative. We first write (8.5) in the form

$$A_k^j(h)\,dh^k = A_k^j(x)\,dx^k \tag{8.5'}$$

where $h = h(z, x)$. This equation is to hold identically in z. Let $u = h(y, z)$, $v = h(z, x)$, $w = h(u, x)$, and $\tilde{w} = h(y, v)$. We regard y and z as fixed and allow x to vary. Then $w = w(x)$ and $\tilde{w} = \tilde{w}(x)$; and we want to show that $w = \tilde{w}$. Now $w(x)$ and $v(x)$ clearly satisfy (8.5'). Therefore

$$A_k^i(\tilde{w})\,d\tilde{w}^k = A_k^i(v)\,dv^k = A_k^i(x)\,dx^k,$$

and so $\tilde{w}$ satisfies (8.5') and hence (8.5). But the solution of (8.5) is uniquely determined by the initial conditions, (see Theorem 8.2 below) and therefore $\tilde{w}(x) = w(x)$, which proves the associativity. Note that associativity is really tied to the left invariance of the Maurer–Cartan forms.

Equations (8.5) comprise an over determined system of differential equations for the functions $h^k(z, x)$. These equations can be solved provided the functions $A_j^i(x)$ satisfy a set of integrability conditions known as the *Maurer* equations:

Theorem 8.2. *Equations* (8.5) *can be solved uniquely for h satisfying the initial conditions $h(z, 0) = z$ and $h(0, x) = x$ if and only if the functions $A_i^j(x)$ satisfy the following integrability conditions (equations of Maurer)*

$$\frac{\partial A_k^i}{\partial x^j} - \frac{\partial A_j^i}{\partial x^k} = \sum_{m,n} \gamma_{mn}^i A_k^m A_j^n \tag{8.7}$$

where the γ_{mn}^i are constants.

The Maurer equations (8.7) are actually the Maurer–Cartan equations in local coordinates, as the reader should check for himself, with the coefficients γ_{mn}^i equal to the structure constants. Equations (8.5) are a special case of the system of A. Mayer:

$$\frac{\partial h^k}{\partial x^l} = F_l^k(x, h) \tag{8.8}$$

where, in this case $F_l^k(x, h) = \sum_j (A^{-1}(h))_j^k A_l^j(x)$.

This is an overdetermined system of partial differential equations for the functions h. Should a smooth solution exist, its mixed partial derivatives must be equal. A necessary condition for solvability is therefore that

$$\frac{\partial^2 h^k}{\partial x^j \partial x^l} = \frac{\partial^2 h^k}{\partial x^l \partial x^j}.$$

Using this and (8.8), we get the equations

$$\frac{\partial F_l^k}{\partial x^j} + \frac{\partial F_l^k}{\partial h^m} F_j^m = \frac{\partial F_j^k}{\partial x^l} + \frac{\partial F_j^k}{\partial x^m} F_l^m.$$

The Maurer equations follow from these after some non-trivial manipulations (see Pontryagin, p. 400).

The classical existence theory of A. Mayer (cf. Caratheodory, p. 26) guarantees that the integrability conditions are also sufficient to prove (at least locally) the existence of a smooth solution. That solution is uniquely determined by the initial conditions $h(z, 0) = z$ and $h(0, x) = x$.

Theorem 8.2 shows that the integrability conditions for the overdetermined system (8.5) are given by the Maurer–Cartan equations (8.6). Our task to construct a Lie group given a set of structure constants is thus reduced to the construction of a set of functions $A_j^i(x)$ such that the one-forms $\omega^i = A_j^i(x)\, dx^j$ satisfy (8.6).

Let $A^i(t, x)$ satisfy the initial value problem

$$\frac{\partial A_j^i}{\partial t} = \delta_j^i - \sum_{k,l} c_{kl}^i x^k A_j^l$$

$$A_j^i(0, x) = 0$$

and put $\omega^i(t, x) = \sum_j A_j^i(t, x)\, dx^j$. We claim $\omega^i(1, x)$ satisfy (8.6). First, since $A_j^i(t, 0) = \delta_j^i t$, it is clear that $\omega^i(1, 0) = dx^i$. Let $\dot{\omega}^i = \partial \omega^i / \partial t$. Then

$$\dot{\omega}^i = dx^i - \sum_{k,l} c_{kl}^i x^k \omega^l$$

and

$$d\dot{\omega}^i = -\sum_{k,l} c_{kl}^i (dx^k \wedge \omega^l + x^k\, d\omega^l).$$

Putting $\theta^i = d\omega^i + \frac{1}{2}\sum_{j,k} c_{jk}^i \omega^j \wedge \omega^k$, we have

$$\dot{\theta}^i = d\dot{\omega}^i + \frac{1}{2}\sum_{j,k} c_{jk}^i (\dot{\omega}^j \wedge \omega^k + \omega^j \wedge \dot{\omega}^k)$$

$$= d\dot{\omega}^i + \sum_{j,k} c_{jk}^i \dot{\omega}^j \wedge \omega^k$$

$$= -\sum_{r,l} x^r c_{rl}^i d\omega^l - \sum_{\substack{k,r,s \\ j}} c_{jk}^i c_{rs}^j x^r \omega^s \wedge \omega^k.$$

From the Jacobi identity and the fact that $\omega^s \wedge \omega^k = -\omega^k \wedge \omega^s$ we derive the identity

$$\sum_{j,s,k} c^i_{jr} c^j_{sk} \omega^s \wedge \omega^k + 2 \sum_{j,k,s} c^i_{jk} c^j_{rs} \omega^s \wedge \omega^k = 0.$$

Therefore

$$\dot\theta^i = -\sum_{r,l} c^i_{rl} x^r \, d\omega^l - \frac{1}{2} \sum_{k,l,r,s} x^r c^i_{rl} c^l_{sk} \omega^s \wedge \omega^k$$

$$= -\sum_{r,l} c^i_{rl} x^r \theta^l.$$

This shows that the coefficients of the two-forms θ^l satisfy a homogeneous linear system of ordinary differential equations (in t). Since $\omega^i(0, x) = 0$, $\theta^i(0, x) = 0$; so by the uniqueness theorem the θ^i vanish identically. This completes the construction of the Maurer–Cartan forms, and thus step (i) of the program. We have now shown that we can construct a (local) Lie group given a set of structure constants which are skew symmetric and satisfy the Jacobi identity.

Recall from Chapter 3 that canonical coordinates of the first kind are co-ordinates such that $x^i(t) = ta^i$, a^i fixed, are the coordinates of a one parameter subgroup. The integration of the Maurer–Cartan equations given here naturally yields canonical coordinates of the first kind (see Pontryagin).

Lie assumed differentiability (at least second order) of the binary composition. Hilbert, in his 5th problem, asked whether this assumption was necessary or whether the differentiable structure of transformation groups might not be a consequence of continuity alone. The affirmative answer to this question was given in 1952 by Gleason, and Montgomery and Zippin.

EXERCISES

1. What are the solvability conditions for the equations $\partial V/\partial x^i = F_i$, $i = 1, 2, 3$? Express the result in terms of differential forms. What are the solvability conditions for the equations curl $\vec{A} = \vec{B}$?

2. Show that the integrability conditions for the Mayer equations can be expressed as $d\omega^i = v$, where $\omega^i = F_j(x, h) \, dx^j$ and $h = (h^1(x), \ldots, h^n(x))$.

ALGEBRAIC THEORY

CHAPTER 9

General Structure of Lie Algebras

30. Ideals, Solvability, and Nilpotency

Recall that in Chapter 2 we introduced the Levi decomposition, according to which every Lie algebra is the semi-direct sum of its radical (the largest solvable ideal) and a semi-simple subalgebra. This reduces the problem of finding all Lie algebras to the three problems of classifying the semi-simple algebras, the solvable algebras, and finding the ways these can fit together in a semi-direct sum. The first problem has been completely resolved by Cartan, and we shall discuss the classification of the semi-simple algebras in the next chapter. The classification of the solvable algebras is an open problem, even though their structure may appear simpler.

The Levi decomposition of the Euclidean group expresses $\mathscr{E}(3)$ as the semi-direct sum of the translations and rotations. The Poincaré group in the theory of relativity is decomposed into the sum of space-time translations and pure Lorentz transformations. O'Raifeartaigh used the Levi decomposition in his proof that the Lorentz group and the internal symmetry groups of elementary particles could not be fit together into a larger group in any but the obvious way as a direct product.

The solvable and semi-simple algebras have a simple characterization in terms of the Killing form. The basic theorems, which will be proved in this chapter, are known as *Cartan's criteria*.

Theorem 9.1. *A Lie algebra $\mathfrak{g}$ is solvable iff $K(X, Y) = 0$ for all $X \in \mathfrak{g}$ and $Y \in \mathfrak{g}^{(1)}$.*

Theorem 9.2. *A Lie algebra $\mathfrak{g}$ is semi-simple iff the Killing form is non-degenerate.*

To these two theorems we may also add the following, whose proof will be given in Chapter 11: *A Lie algebra* $\mathfrak{g}$ *generates a compact group iff its Killing form is negative definite.* Thus, an algebraic property of $\mathfrak{g}$ enforces a topological property on its group.

In this section we summarize a number of preliminary algebraic facts needed in the sequel. We begin with a proof of the invariance properties (2.5) and (2.6) of the Killing form.

Property (2.5) follows from the fact that if ρ is an automorphism, then $\mathrm{ad}\,\rho(X) = \rho\,\mathrm{ad}\,X\rho^{-1}$; and therefore

$$
\begin{aligned}
K(\rho(X), \rho(Y)) &= \mathrm{Tr}\,\mathrm{ad}\,\rho(X)\mathrm{ad}\,\rho(Y) \\
&= \mathrm{Tr}\,\rho\,\mathrm{ad}\,X\rho^{-1}\rho\,\mathrm{ad}\,Y\rho^{-1} \\
&= \mathrm{Tr}\,\rho\,\mathrm{ad}\,X\,\mathrm{ad}\,Y\rho^{-1} \\
&= \mathrm{Tr}\,\mathrm{ad}\,X\,\mathrm{ad}\,Y \\
&= K(X, Y).
\end{aligned}
$$

Equation (2.6) is the infinitesimal version of (2.5). Let ρ_t be the one-parameter group of automorphisms of $\mathfrak{g}$ given by $\rho_t Y = e^{tz} Y e^{-tz}$. By (2.5) $K(\rho_t X, \rho_t Y) = K(X, Y)$. Differentiating with respect to t and setting $t = 0$ we get (2.6).

We may also prove (2.6) in the following way. From the Jacobi identity,

$$
\begin{aligned}
K(X, [Y, Z]) &= \mathrm{Tr}(\mathrm{ad}\,X\,\mathrm{ad}[Y, Z]) \\
&= \mathrm{Tr}(\mathrm{ad}\,X[\mathrm{ad}\,Y, \mathrm{ad}\,Z]) \\
&= \mathrm{Tr}(\mathrm{ad}\,Y\,\mathrm{ad}\,Z\,\mathrm{ad}\,X - \mathrm{ad}\,Y\,\mathrm{ad}\,X\,\mathrm{ad}\,Z) \\
&= \mathrm{Tr}(\mathrm{ad}\,Y[\mathrm{ad}\,Z, \mathrm{ad}\,X]) \\
&= K(Y, [Z, X]).
\end{aligned}
$$

(Recall that $\mathrm{Tr}\,ABC$ is invariant under cyclic permutations of A, B, and C.)

Let $\{X_1, \ldots, X_n\}$ be a basis for $\mathfrak{g}$. The components of the Killing form in this basis are

$$
g_{rs} = K(X_r, X_s) = \sum_{j,k=1}^{n} C_{rj}^{k} C_{sk}^{j}.
$$

If we make a transformation to a new basis given by $X'_i = \sum_j O_{ij} X_j$ then the metric tensor is transformed into $g'_{rs} = \sum_{j,k} O_{rj} g_{jk} O_{sk}$; that is, $g' = OgO^t$. Now g is symmetric. If g is also real then there is an orthogonal matrix O which diagonalizes g. There exists therefore a basis $\{X_1, \ldots, X_n\}$ in which the metric tensor is diagonal.

An instructive calculation involving the metric tensor is in the proof of the following lemma.

Lemma 9.3. *If* $\mathfrak{a}$ *is an ideal in* $\mathfrak{g}$, *the Killing form of* $\mathfrak{a}$ *(considered as a Lie algebra in its own right) is the Killing form of* $\mathfrak{g}$ *restricted to* $\mathfrak{a}$: $K_{\mathfrak{a}}(X, Y) = K_{\mathfrak{g}}(X, Y)$ *for all* $X, Y \in \mathfrak{a}$.

PROOF. Let $\{X_1, \ldots, X_n\}$ be a basis for $\mathfrak{g}$ such that $\{X_1, \ldots, X_r\}$ is a basis for $\mathfrak{a}$. Since $[\mathfrak{a}, \mathfrak{g}] \subset \mathfrak{a}$ the structure constants $C_{ij}^k = 0$ when $i \leq r < k$ and all j. So for $1 \leq i, j \leq r$,

$$K_\mathfrak{a}(X_i, X_j) = \operatorname{Tr} \operatorname{ad} X_i \operatorname{ad} X_j$$

$$= \sum_{k, l=1}^{r} C_{il}^k C_{jk}^l$$

$$= \sum_{k, l=1}^{n} C_{il}^k C_{jk}^l$$

$$= (K_\mathfrak{g})_{ij}. \qquad \square$$

We now discuss a number of algebraic results concerning ideals of a Lie algebra. If $\mathfrak{a}$ and $\mathfrak{b}$ are ideals then $\mathfrak{a} \cap \mathfrak{b}$, $\mathfrak{a} + \mathfrak{b} = \{X + Y | X \in \mathfrak{a}, Y \in \mathfrak{b}\}$ and $[\mathfrak{a}, \mathfrak{b}]$ are ideals. The kernel of a homomorphism is an ideal. The adjoint representation $X \to \operatorname{ad} X$ is a homomorphism of $\mathfrak{g}$ into the linear transformations on $\mathfrak{g}$; its kernel is the center $\mathfrak{Z}(\mathfrak{g}) = \{X | [X, Y] = 0 \text{ for all } Y \in \mathfrak{g}\}$. The upper central series, defined by $\mathfrak{g}^{(0)} = \mathfrak{g}$, $\mathfrak{g}^{(j+1)} = [\mathfrak{g}^{(j)}, \mathfrak{g}^{(j)}]$ is a sequence of ideals, as is the lower central series: $\mathfrak{g}_{(0)} = \mathfrak{g}$, $\mathfrak{g}_{(j+1)} = [\mathfrak{g}, \mathfrak{g}_{(j)}]$.

Recall that $\mathfrak{g}$ is solvable if $\mathfrak{g}^{(n)} = 0$ for some n and $\mathfrak{g}$ is nilpotent if $\mathfrak{g}_{(n)} = 0$ for some n. A nilpotent algebra is always solvable, but not conversely.

If ρ is a homomorphism of $\mathfrak{g}$ into an algebra $\mathfrak{g}'$ then $\ker \rho$ is an ideal in $\mathfrak{g}$ and $\rho(\mathfrak{g})$ is isomorphic to $\mathfrak{g}/\ker \rho$.

It is easily seen that $(\rho(\mathfrak{g}))^{(j)} = \rho(\mathfrak{g}^{(j)})$; hence the homomorphic image of a solvable algebra is solvable. The same holds for nilpotent algebras. It is clear that subalgebras of solvable and nilpotent algebras are themselves solvable and nilpotent respectively. Moreover

Lemma 9.4. *If $\mathfrak{a}$ is a solvable ideal of $\mathfrak{g}$ and if $\mathfrak{g}/\mathfrak{a}$ is solvable, then $\mathfrak{g}$ is solvable. If $\mathfrak{a}$ is a nilpotent ideal contained in the center of $\mathfrak{g}$ and $\mathfrak{g}/\mathfrak{a}$ is nilpotent, then $\mathfrak{g}$ is nilpotent.*

PROOF. $(\mathfrak{g}/\mathfrak{a})^{(j)} = \mathfrak{g}^{(j)}/\mathfrak{a}$; so if $\mathfrak{g}/\mathfrak{a}$ is solvable then $\mathfrak{g}^{(n)} \subseteq \mathfrak{a}$ for some n. But then $\mathfrak{g}^{(n)}$ is solvable since $\mathfrak{a}$ is solvable; and therefore $\mathfrak{g}$ is solvable. The corresponding argument for nilpotent algebras is based on the identity $(\mathfrak{g}/\mathfrak{a})_{(j)} = \mathfrak{g}_{(j)}/\mathfrak{a}$. If $\mathfrak{g}/\mathfrak{a}$ is nilpotent then $\mathfrak{g}_{(j)}/\mathfrak{a} = \mathfrak{a}$ for some j, hence $\mathfrak{g}_{(j)} \subseteq \mathfrak{a}$. In that case $\mathfrak{g}_{(j+1)} = [\mathfrak{g}, \mathfrak{g}_{(j)}] \subseteq [\mathfrak{g}, \mathfrak{a}] = 0$ since $\mathfrak{a}$ is contained in the center of $\mathfrak{g}$. Hence $\mathfrak{g}$ is nilpotent. $\qquad \square$

We may apply this to prove the following:

Lemma 9.5. *Let ρ be a homomorphism of $\mathfrak{g}$; if $\rho(\mathfrak{g})$ and $\ker \rho$ are both solvable then so is $\mathfrak{g}$. In particular $\mathfrak{g}$ is solvable iff $\operatorname{ad} \mathfrak{g}$ is solvable. Moreover, $\mathfrak{g}$ is nilpotent if $\operatorname{ad} \mathfrak{g}$ is nilpotent.* (By $\operatorname{ad} \mathfrak{g}$ we mean the image of $\mathfrak{g}$ under the adjoint representation, that is, the matrices of the adjoint representation.)

PROOF. By the first homomorphism theorem, which is a general result for homomorphisms of rings, $\mathfrak{g}/\ker \rho$ is isomorphic to $\rho(\mathfrak{g})$, so the first statement follows from Lemma 9.4. The second statement follows from the first by noting that the adjoint representation is a Lie algebra homomorphism and that the kernel of this homomorphism is the center of the algebra, which is abelian, hence both solvable and nilpotent. $\qquad\square$

Theorem 9.6. *The sum of two solvable ideals is a solvable ideal. Consequently, every Lie algebra has a unique radical, obtained as the sum of all its solvable ideals.*

PROOF. If $\mathfrak{a}$ and $\mathfrak{b}$ are solvable ideals then $\mathfrak{a} \cap \mathfrak{b}$ is an ideal in $\mathfrak{a}$, hence is solvable. Furthermore, $[\mathfrak{a} + \mathfrak{b}, \mathfrak{g}] = [\mathfrak{a}, \mathfrak{g}] + [\mathfrak{b}, \mathfrak{g}] \subset \mathfrak{a} + \mathfrak{b}$, so $\mathfrak{a} + \mathfrak{b}$ is an ideal. By the third isomorphism theorem (see exercise 10), $(\mathfrak{a} + \mathfrak{b})/\mathfrak{b} \cong \mathfrak{a}/(\mathfrak{a} \cap \mathfrak{b})$. But $\mathfrak{a}/(\mathfrak{a} \cap \mathfrak{b})$ is a homomorphic image of $\mathfrak{a}$, so is solvable. Therefore $\mathfrak{a} + \mathfrak{b}$ is solvable by Lemma 9.4. $\qquad\square$

An immediate consequence of the first statement is that every finite dimensional Lie algebra $\mathfrak{g}$ possesses a unique largest solvable ideal $\mathfrak{R}$, called the radical, obtained as the sum of all the solvable ideals of $\mathfrak{g}$. It is easily seen that $\mathfrak{g}/\mathfrak{R}$ is semi-simple. The Levi decomposition states that $\mathfrak{g}$ actually contains a semi-simple subalgebra $\mathfrak{S}$ such that $\mathfrak{g} = \mathfrak{S} \oplus_s \mathfrak{R}$ (see Jacobson).

EXERCISES

1. Prove that $(\rho(\mathfrak{g}))^{(j)} = \rho(\mathfrak{g}^{(j)})$, where ρ is a homomorphism of $\mathfrak{g}$.

2. Let ρ be a representation of a Lie algebra $\mathfrak{g}$ by linear operators on a vector space and define $K_\rho(X, Y) = \operatorname{Tr} \rho(X)\rho(Y)$. Show that $K_\rho([X, Y], Z) + K_\rho(Y, [X, Z]) = 0$.

3. The radical of the Killing form is the set $\operatorname{rad} K = \{X | K(X, Y) = 0 \text{ for all } Y \in \mathfrak{g}\}$. Show that $\operatorname{rad} K$ is a solvable ideal of $\mathfrak{g}$. Consider the Lie algebra $\{A, B\}$ where $[A, B] = B$. What is the radical of the algebra? Of the Killing form?

4. If $\mathfrak{a}$ is an ideal then $\mathfrak{a}^\perp = \{X | K(X, Y) = 0 \text{ for all } Y \in \mathfrak{a}\}$ is an ideal. Show that a semi-simple algebra can be decomposed into a direct sum of orthogonal ideals. Show $so(4) \cong so(3) \oplus so(3)$.

5. The derivations of a Lie algebra $\mathfrak{g}$ themselves form a Lie algebra. The *inner derivations* of $\mathfrak{g}$ are those of the form $DZ = (\operatorname{ad} X)(Z)$ for some $X \in \mathfrak{g}$. Show that the inner derivations form an ideal.

6. Show $\mathfrak{g}/\mathfrak{g}^{(1)}$ is abelian. What is $[e(3), e(3)]$? If $\mathfrak{a}$ is an ideal so are all the $\mathfrak{a}^{(j)}$ and $\mathfrak{a}_{(j)}$.

7. $sl(n, c)$ is an ideal of $\mathfrak{gl}(n, c)$; $o(n, c)$ is a subalgebra but not an ideal of $\mathfrak{gl}(n, c)$.

8. Let K be a non-degenerate bilinear form on a vector space V and let W be a subspace of V. Prove that $\dim W + \dim W^\perp = \dim V$. (*Hint:* Let $v_1, \dots, v_p$ be a basis for W and consider the linear transformation $f: V \to \mathbb{R}^n$ given by $f(w) = (K(v_1, w), \dots, K(v_p, w))$. Then $\operatorname{rank} f + \operatorname{nullity} f = \dim v$.)

9. Give an example of a solvable algebra which is not nilpotent.

10. Let $\mathfrak{a}, \mathfrak{b}$ be ideals in a Lie algebra $\mathfrak{g}$. Show that $\mathfrak{a}$ and $\mathfrak{b}$ are ideals in $\mathfrak{a} + \mathfrak{b}$ and that $\mathfrak{a} \cap \mathfrak{b}$ is an ideal in $\mathfrak{a}$. Prove that $(\mathfrak{a} + \mathfrak{b})/\mathfrak{a} \cong \mathfrak{a}/\mathfrak{a} \cap \mathfrak{b}$. (*Hint*: $(\mathfrak{a} + \mathfrak{b})/\mathfrak{b}$ is the set of cosets $X + \mathfrak{b}$, with $X \in \mathfrak{a}$; consider the homomorphism from $\mathfrak{a}$ to $(\mathfrak{a} + \mathfrak{b})/\mathfrak{b}$ given by $X \to X + \mathfrak{b}$ for $X \in \mathfrak{a}$.)

31. Theorems of Lie and Engels

Let $\mathfrak{N}_k$ be the set of complex upper triangular $n \times n$ matrices such that $a_{ij} = 0$ for $j < k + i$. From the commutation relations $[\mathfrak{N}_k, \mathfrak{N}_l] \subseteq \mathfrak{N}_{k+l}$ it follows that $\mathfrak{N}_0$ is solvable and $\mathfrak{N}_1$ is nilpotent. The algebra $\lambda I + \mathfrak{N}_1$, where I is the identity matrix, is also nilpotent, as are the direct sums of such algebras, given by

where $m_1, \ldots, m_r$ are non-negative integers and $\lambda_1, \ldots, \lambda_r$ are arbitrary complex numbers. We denote the above direct sum by $S(m_1, \ldots, m_r)$.

We shall see, from the theorems of Lie and Engels, that the complex solvable matrix algebras are isomorphic to $\mathfrak{N}_0$ and its subalgebras, and that the nilpotent algebras are subalgebras of $S(m_1, \ldots, m_r)$. Before proceeding to the statements and proofs of those theorems we summarize some important facts from linear algebra.

A linear transformation $X \in gl(V)$ is said to be *diagonalizeable*, or *semi-simple*, if there exists a basis of V consisting entirely of eigenvectors of X. In that case the matrix of X with respect to this basis is a diagonal matrix. A family of operators $\mathfrak{F} \subset gl(V)$ is said to be *simultaneously diagonalizeable* if there is a basis for V which diagonalizes all of them at the same time. A necessary and sufficient condition for a family of semi-simple operators to be simultaneously diagonalizeable is that they all commute—i.e. form an abelian algebra.

Recall that in quantum mechanics observables are represented by Hermitian operators $A, B, \ldots$. A commuting set of operators represents a set of simultaneous observables. For example, in the theory of the hydrogen atom the Hamiltonian H, the total angular momentum J^2 and the azimuthal angular momentum J_z all commute, so the respective quantities can be measured simultaneously without any *a priori* restriction on the accuracy of the measurements.

Since L_x, and L_y do not commute with L_z, however, the three components of angular momentum can not be measured simultaneously: the accuracy of any simultaneous measurement is fundamentally limited by the Heisenberg uncertainty principle.

The same applies to the position and momentum of a particle. Since the position and momentum operators satisfy the commutation relation $[P, Q] = i\hbar\mathbb{1}$, there is a fundamental limit to the accuracy of simultaneous measurement of the position and momentum of a particle. In this case it is given by

$$(\Delta P)(\Delta Q) \geq \tfrac{1}{2}\hbar.$$

This is the Heisenberg uncertainty principle. (For a mathematical statement and proof, see Weyl [3].)

If V is a complex vector space then every $X \in gl(V)$ may be represented by an upper triangular matrix (for an appropriate choice of basis). If the matrix of X with respect to the basis $\{v_1, \ldots, v_n\}$ is upper triangular then Xv_i is a linear combination of $v_1, \ldots, v_i$. We may express this by writing

$$Xv_i \equiv 0 \qquad (\mathrm{mod}\, v_1, \ldots, v_i);$$

or, we may put $V_i = \mathrm{span}[v_1, \ldots, v_i]$ and express this condition as $XV_i \subseteq V_i$.

We say X is strictly upper triangular if in addition the diagonal entries of its matrix are zero. In that case

$$Xv_i \equiv 0 \qquad (\mathrm{mod}\, v_1, \ldots, v_{i-1})$$

or $XV_i \subseteq V_{i-1}$. If X is strictly upper triangular then $X^k = 0$ for some k and X is nilpotent. On the other hand, if X is nilpotent set $W_i = X^i V$. Then $XW_i \subset W_{i+1}$ and $W_k = 0$ for some k.

Theorem 9.7 (Jordan Decomposition). *Let $X \in gl(V)$, V a complex vector space. Then X can be uniquely decomposed into the sum $X = S + N$ where S is semi-simple and N is nilpotent; $[S, N] = 0$; and S and N are polynomials in X.*

(For a proof, see Hoffmann and Kunze, p. 222.) In fact, there is a basis for V relative to which the matrix of X is

$$
\begin{pmatrix}
\lambda_1 & 1 & 0 & \cdots & & & & & & \\
 & \lambda_1 & 1 & & & & & & & \\
 & & \ddots & 1 & m_1 & & & & & \\
0 & & & \lambda_1 & & & & & & \\
\hline
 & m_1 & & & \lambda_2 & 1 & 0 & \cdots & 0 & \\
 & & & & 0 & \lambda_1 & 1 & \cdots & 0 & \\
 & & & & & & \ddots & & & m_2 \\
 & & & & & & & & 1 & \\
 & & & & 0 & & \cdots & & \lambda_2 & \\
\hline
 & & & & & m_2 & & & & \ddots
\end{pmatrix}
$$

A family $\mathfrak{F} \subset gl(V)$ of triangulable operators is said to be simultaneously triangulable if there exists a basis of V relative to which all elements of $\mathfrak{F}$ have upper triangular matrices. Lie's theorem gives a necessary and sufficient condition for the elements of a Lie algebra to be simultaneously upper triangular:

Theorem 9.8 (Lie). *Let $\mathfrak{g}$ be a subalgebra of $gl(V)$, V a complex vector space. Then the elements of $\mathfrak{g}$ are simultaneously triangulable iff $\mathfrak{g}$ is solvable.*

The proof will be given below. Lie's theorem may be viewed as a generalization of the theorem that an algebra of semi-simple operators is simultaneously diagonalizeable iff it is abelian.

If $\mathfrak{g}$ is nilpotent then $\mathfrak{g}_{(n)} = 0$ for some n and ad $X_1 \ldots$ ad $X_n = 0$ for any n elements $X_1, \ldots, X_n$ in $\mathfrak{g}$. In particular $(\text{ad } X)^n = 0$ for every X in $\mathfrak{g}$. Such elements are said to be *ad-nilpotent*, and so every element of a nilpotent algebra is ad-nilpotent. On the other hand, the algebra $\{\lambda\mathbb{1} \,|\, \lambda \in C\}$ is a nilpotent algebra none of whose elements are themselves nilpotent. Engel's theorem states:

Theorem 9.9 (Engel). *A Lie algebra $\mathfrak{g}$ is nilpotent iff every element of $\mathfrak{g}$ is ad-nilpotent. Every complex nilpotent linear algebra is isomorphic to a subalgebra of $S(m_1 \ldots m_r)$ for some choice of $m_1, \ldots, m_r$.*

We begin with a proof of Engel's theorem.

Lemma 9.10. *Every nilpotent element in $gl(V)$ is ad-nilpotent.*

PROOF. (Cf. exercise 1.)   □

Lemma 9.11. *Let $\mathfrak{a}$ be an ideal in $\mathfrak{g} \subset gl(V)$. Let $W = \{v \in V \,|\, Xv = 0$ for all $X \in \mathfrak{a}\}$. Then W is invariant under $\mathfrak{g}$.*

PROOF. Exercise.   □

Lemma 9.12. *Let $\mathfrak{g}$ be a subalgebra of $gl(V)$ all of whose elements are nilpotent. Then there is a $v \in V$ such that $v \neq 0$ and $Xv = 0$ for all $X \in \mathfrak{g}$.*

PROOF. The result is trivial if dim $\mathfrak{g} = 1$; we proceed by induction on dim $\mathfrak{g}$. Suppose the result is true for any algebra of dimension less than r, let $\mathfrak{g}$ have dimension r, and let $\mathfrak{a}$ be a proper subalgebra of $\mathfrak{g}$ of maximum dimension. Let us show that $\mathfrak{a}$ is an ideal of codimension 1. Consider the action of $\mathfrak{a}$ on the coset space $\mathfrak{g}/\mathfrak{a}$ defined by

$$X(Y + \mathfrak{a}) = \text{ad } X(Y) + \mathfrak{a}$$

for each $X \in \mathfrak{a}$ and $Y \in \mathfrak{g}$. The operators ad X form a Lie algebra of operators acting on the cosets of $\mathfrak{g}/\mathfrak{a}$. Obviously the dimension of this algebra is not greater than that of $\mathfrak{a}$. Moreover, each operator ad X is nilpotent by Lemma 9.10. Therefore by the induction hypothesis there is an element $S + \mathfrak{a}$ in $\mathfrak{g}/\mathfrak{a}$ such that $S + \mathfrak{a} \neq \mathfrak{a}$ yet $X(S + \mathfrak{a}) = \mathfrak{a}$ for all $X \in \mathfrak{a}$; i.e. $S \notin \mathfrak{a}$ but $[X, S] \in \mathfrak{a}$ for all $X \in \mathfrak{a}$. Therefore $[S] + \mathfrak{a}$ is a subalgebra of $\mathfrak{g}$ which strictly contains $\mathfrak{a}$. By the maximality of dim $\mathfrak{a}$, $\mathfrak{g} = [S] + \mathfrak{a}$, and it follows that $\mathfrak{a}$ is an ideal.

Now put $W = \{v \in V | v = 0\}$. Then W is invariant under $\mathfrak{g}$ by Lemma 9.11. Since S is nilpotent and leaves W invariant, there is a vector $v \in W$ for which $Sv = 0$. This v provides the required vector of Lemma 9.12.   □

Lemma 9.13. *Any Lie algebra $\mathfrak{g}$ of nilpotent transformations is isomorphic to a subalgebra of $\mathfrak{N}_1$, hence is a nilpotent algebra.*

PROOF. By Lemma 9.12 there exists a v_1 such that $Xv_1 = 0$ for all $X \in \mathfrak{g}$. We define the action of $\mathfrak{g}$ on the quotient space $V/[v_1]$ by

$$X(w + [v_1]) = Xw + [v_1].$$

Applying the lemma to $\mathfrak{g}$ acting on $V/[v_1]$ we see there exists a v_2 such that $X(v_2 + [v_1]) = [v_1]$, i.e. $Xv_2 \equiv 0 \pmod{v_1}$. We then apply the lemma to $\mathfrak{g}$ acting on $V/[v_1, v_2]$, and so forth. In the end we arrive at a *composition series* $V_i = sp[v_1 \ldots v_i]$,

$$0 \subset V_1 \subset V_2 \subset \cdots \subset V_n = V,$$

$$\dim V_i/V_{i-1} = 1 \qquad \mathfrak{g}V_{i+1} \subset V_i.$$

The matrices of all transformations in $\mathfrak{g}$ with respect to this basis therefore all lie in $\mathfrak{N}_1$; and so $\mathfrak{g}$ is isomorphic to a subalgebra of $\mathfrak{N}_1$, which is nilpotent. This proves Lemma 9.13.   □

Engel's theorem now follows immediately. For if all the elements of $\mathfrak{g}$ are ad-nilpotent, then ad $\mathfrak{g}$ is nilpotent; so by Lemma 9.5, $\mathfrak{g}$ is nilpotent. The verification that $\mathfrak{g}$ is isomorphic to a subalgebra of $S(m_1 \ldots m_r)$ is left to the reader.

We now turn to the proof of Lie's theorem, beginning with

Lemma 9.14. *Let $\mathfrak{g}$ be a solvable subalgebra of $gl(V)$, V a complex vector space. There exists a vector $v \in V$ which is a simultaneous eigenvector of all $X \in \mathfrak{g}$. Thus, there exists a linear functional λ on $\mathfrak{g}$ such that $Xv = \lambda(X)v$ for all $X \in \mathfrak{g}$.*

PROOF. We induct on dim $\mathfrak{g}$, the result being trivial when dim $\mathfrak{g} = 1$. Since $\mathfrak{g}$ is solvable it properly contains $\mathfrak{g}^{(1)}$. Let $\mathfrak{a}$ be a subspace of $\mathfrak{g}$ of codimension 1 which contains $\mathfrak{g}^{(1)}$. Then $\mathfrak{a}$ is an ideal, since $[\mathfrak{a}, \mathfrak{g}] \subset [\mathfrak{g}, \mathfrak{g}] = \mathfrak{g}^{(1)} \subset \mathfrak{a}$. Since $\mathfrak{g}$ is solvable, $\mathfrak{a}$ is solvable. Therefore by our induction hypothesis there exists a vector $v_0 \in V$ which is a common eigenvector of all $X \in \mathfrak{a}$. We may write $Xv_0 = \lambda(X)v_0$ where λ is a linear functional on $\mathfrak{a}$. Choose $Z \in \mathfrak{g} \backslash \mathfrak{a}$ and put $v_{j+1} = Zv_j$ for $j = 0, 1, 2, \ldots$. The subspace W spanned by these vectors must be spanned by some initial sequence of them; say $W = sp[v_1, \ldots, v_p]$. Therefore W is a subspace of V which is invariant under Z. For $X \in \mathfrak{a}$,

$$Xv_1 = XZv_0 = ZXv_0 + [X,Z]v_0$$
$$= \lambda(X)v_1 + \lambda([X,Z])v_0.$$

($[X,Z] \in \mathfrak{a}$ since $\mathfrak{a}$ is an ideal.) By induction we find that $Xv_j \equiv \lambda(X)v_j$ (mod $v_0, \ldots, v_{j-1}$) for all $X \in \mathfrak{a}$. Therefore W is invariant under $\mathfrak{a}$. Furthermore, the matrices of all $X \in \mathfrak{a}$ relative to the basis $v_0, \ldots, v_p$ are triangular; that is, they take the form $X = \lambda(X)I + N(X)$ where N is nilpotent. From this it follows that $\operatorname{Tr}_W(X) = \lambda(X) \dim W$. Since $\operatorname{Tr}_W([X,Z]) = 0$, it follows that $\lambda([X,Z]) = 0$ for any $X \in \mathfrak{a}$. Now

$$Xv_{j+1} = XZv_j = ZXv_j + [X,Z]v_j$$
$$= \lambda(X)v_{j+1} + [X,Z]v_j.$$

Since $Xv_0 = \lambda(X)v_0$, we find, by induction on j, that $Xv_j = \lambda(X)v_j$ for all j and all $X \in \mathfrak{a}$. Thus the elements of $\mathfrak{a}$ are simultaneously diagonal with respect to the basis $v_0, v_1 \ldots v_p$. Since Z leaves W invariant it has an eigenvector $w \in W$; w is the required eigenvector.

We are now ready to prove Lie's theorem. Let v_1 be a common eigenvector of all $X \in \mathfrak{g}$ and let $Q_1 = V/[v_1]$. Since all transformations in $\mathfrak{g}$ leave v_1 invariant they can be carried to the quotient space. Applying the lemma to $\mathfrak{g}$ acting on Q_1, we see there is a vector v_2 such that the coset $v_2 + [v_1]$ is a common eigenvector of all $X \in \mathfrak{g}$; i.e. $X(v_2 + [v_1]) = Xv_2 + [v_1] = \lambda_2(X)v_2 + [v_1]$. Continuing in this manner we find a sequence of vectors $v_1 \ldots v_n$ of V such that $Xv_i \equiv \lambda_i(X)v_i$ (mod $v_1, \ldots, v_{i-1}$). This means that all elements of $\mathfrak{g}$ are represented by upper triangular matrices relative to the basis $\{v_1, \ldots, v_n\}$. $\square$

EXERCISES

1. If X is nilpotent then $\operatorname{ad} X$ is nilpotent. (*Hint*: set $L_X Y = XY$ and $R_X Y = YX$. Then $\operatorname{ad} X = L_X - R_X$. What is $(\operatorname{ad} X)^k$?)

2. Prove Lemma 9.11.

3. If $\mathfrak{g} \subset gl(V)$ is solvable there is a composition series $0 = V_0 \subset V_1 \subset \cdots \subset V_n = V$ such that $\mathfrak{g}V_i \subset V_i$ and $\dim V_i = i$. If $\mathfrak{g}$ is solvable there is a sequence of ideals $\mathfrak{g}_i$ in $\mathfrak{g}$ such that $0 = \mathfrak{g}_0 \subset \mathfrak{g}_1 \subset \cdots \subset \mathfrak{g}_n = \mathfrak{g}$, $\dim \mathfrak{g}_i = i$, and $[\mathfrak{g}, \mathfrak{g}_i] \subset \mathfrak{g}_i$.

4. $\mathfrak{g}$ is solvable iff $\mathfrak{g}^{(1)}$ is nilpotent. If $\mathfrak{g}$ is nilpotent there exists a chain of ideals $\mathfrak{g}_i$ such that $\dim \mathfrak{g}_i = i$, $0 = \mathfrak{g}_0 \subset \mathfrak{g}_1 \subset \cdots \subset \mathfrak{g}_n = \mathfrak{g}$ and $[\mathfrak{g}, \mathfrak{g}_i] \subset \mathfrak{g}_{i-1}$.

32. Cartan's Criteria

We are now ready to give the proofs of Cartan's criteria, Theorems 9.1 and 9.2. We first need

Theorem 9.15. *Let* $\mathfrak{g}$ *be a subalgebra of* $gl(V)$ *and suppose that* $\operatorname{Tr} XY = 0$ *for all* $X, Y \in \mathfrak{g}$. *Then* $\mathfrak{g}$ *is solvable.*

PROOF. We need not assume here that $\mathfrak{g}$ and V are complex. If they are real we may complexify them. The condition $\operatorname{Tr} XY = 0$ still holds in the complexified algebra, and if the complexified algebra is solvable so is $\mathfrak{g}$. Thus it suffices to prove the theorem for complex algebras and vector spaces.

By exercise 4 of §31, it suffices to show that $\mathfrak{g}^{(1)}$ is nilpotent, and by Lemma 9.13 this may be done by showing that each $X \in \mathfrak{g}^{(1)}$ is nilpotent. For $X \in \mathfrak{g}^{(1)}$ let $X = S + N$ be its Jordan decomposition, with $S = \operatorname{diag}(\lambda_1, \ldots, \lambda_n)$. Now $\operatorname{ad} S$ is diagonal; in fact, $\operatorname{ad} S(E_{ij}) = (\lambda_i - \lambda_j)E_{ij}$, where E_{ij} is the matrix with a 1 in the ij^{th} entry and zeros elsewhere. Furthermore, $\operatorname{ad} N$ is nilpotent since N is nilpotent. The Jordan decomposition being unique, $\operatorname{ad} S + \operatorname{ad} N$ is the Jordan decomposition of $\operatorname{ad} X$, and consequently $\operatorname{ad} S$ is a polynomial in $\operatorname{ad} X$. We may write $\operatorname{ad} S = p(\operatorname{ad} X)$; then for any $Y \in \mathfrak{g}$, $[S, Y] = \operatorname{ad} S(Y) = p(\operatorname{ad} X)(Y) \in \mathfrak{g}$. Let $\bar{S} = \operatorname{diag}(\bar{\lambda}_1, \ldots, \bar{\lambda}_n)$. Then $[\operatorname{ad} \bar{S}, \operatorname{ad} S] = \operatorname{ad}[\bar{S}, S] = 0$; so $\operatorname{ad} \bar{S}$ is a polynomial in $\operatorname{ad} S$ and therefore in $\operatorname{ad} X$. It follows that $[\bar{S}, Y] \in \mathfrak{g}$ as well. Moreover since $\bar{S}$ is diagonal, $[\bar{S}, N] = 0$, so $\bar{S}N$ is nilpotent. It follows that $\operatorname{Tr} \bar{S}X = \operatorname{Tr} \bar{S}(S + N) = \operatorname{Tr} \bar{S}S = \sum_i |\lambda_i|^2$. On the other hand, since $X \in \mathfrak{g}^{(1)}$ it can be expressed in the form $X = \sum_r [A_r, B_r]$; hence $\operatorname{Tr} \bar{S}X = \sum_r \operatorname{Tr} \bar{S}[A_r, B_r] = \sum_r \operatorname{Tr}[\bar{S}, A_r]B_r$. But each term in this sum vanishes by the hypotheses of the theorem, since $[\bar{S}, A_r] \in \mathfrak{g}$. Therefore $\operatorname{Tr} \bar{S}X = 0$, $\sum_i |\lambda_i|^2 = 0$, and $S = 0$. Consequently X is nilpotent and we have shown that every $X \in \mathfrak{g}^{(1)}$ is nilpotent. Accordingly $\mathfrak{g}^{(1)}$ is a nilpotent algebra, which is what we needed to prove. □

The proof of Theorem 9.1 is now very simple. If $K(X, Y) = 0$ for all $Y \in \mathfrak{g}^{(1)}$ then certainly $K(X, Y) = 0$ when both X and Y are in $\mathfrak{g}^{(1)}$. Thus $\operatorname{Tr} \operatorname{ad} X \operatorname{ad} Y = 0$ for all X and Y in $\mathfrak{g}^{(1)}$ and so $\operatorname{ad} \mathfrak{g}^{(1)}$ is solvable by Theorem 9.15; $\mathfrak{g}^{(1)}$ is solvable by Lemma 9.5; and so $\mathfrak{g}$ itself is solvable.

Conversely, suppose $\mathfrak{g}$ is solvable. If $\mathfrak{g}$ is real we may complexify it, and its complexification is also solvable. Therefore $\operatorname{ad} \mathfrak{g}$ is solvable, and by Lie's theorem it is isomorphic to some subalgebra $\mathfrak{N}_0$ of upper triangular matrices. It follows that $\operatorname{ad} \mathfrak{g}^{(1)} = (\operatorname{ad} \mathfrak{g})^{(1)} \subset \mathfrak{N}_1$; and so $\operatorname{ad} Y$ is strictly upper triangular for any $Y \in \mathfrak{g}^{(1)}$. But then $\operatorname{ad} X \operatorname{ad} Y \in \mathfrak{N}_1$ for all $X \in \mathfrak{g}$ and $Y \in \mathfrak{g}^{(1)}$, and so $K(X, Y) = 0$ for all $X \in \mathfrak{g}$ and $Y \in \mathfrak{g}^{(1)}$.

Now let us prove Theorem 9.2. If $\mathfrak{g}$ is not semi-simple it has an abelian ideal $\mathfrak{a}$. For $X \in \mathfrak{a}$ and $Y \in \mathfrak{g}$ it is easily seen that $(\operatorname{ad} X \operatorname{ad} Y)^2 = 0$; hence $\operatorname{ad} X \operatorname{ad} Y$ is nilpotent and its trace is zero. Thus $K(X, Y) = 0$ for all $X \in \mathfrak{a}$ and K is degenerate.

Conversely, if K is degenerate then $\operatorname{rad} K = \{X | K(X, Y) = 0 \text{ for all } Y \in \mathfrak{g}\}$ is an ideal. Now K restricted to $\operatorname{rad} K$ vanishes identically by Lemma 9.3; so by Theorem 9.1 $\operatorname{rad} K$ is a solvable ideal and $\mathfrak{g}$ cannot be semi-simple.

We conclude this section with two basic theorems about the structure of semi-simple Lie algebras.

Theorem 9.16. *A semi-simple algebra can be decomposed into a direct sum of orthogonal simple ideals.*

PROOF. If $\mathfrak{g}$ is simple we are done. Otherwise let $\mathfrak{a}$ be an ideal of $\mathfrak{g}$ and let $\mathfrak{a}^\perp$ be its orthogonal complement (relative to the Killing form). Recall that $\mathfrak{a}^\perp$ is an ideal. Since K is non-degenerate $\dim \mathfrak{a} + \dim \mathfrak{a}^\perp = \dim \mathfrak{g}$ (see exercise 8, §30.) Furthermore, $\mathfrak{a} \cap \mathfrak{a}^\perp$ is an abelian ideal (why?); hence $\mathfrak{a} \cap \mathfrak{a}^\perp = \{0\}$. Therefore $\mathfrak{g} = \mathfrak{a} \oplus \mathfrak{a}^\perp$. The argument is now repeated on $\mathfrak{a}$ and $\mathfrak{a}^\perp$, if necessary; and the process is continued until the desired decomposition is attained. ☐

A *derivation* on a Lie algebra $\mathfrak{g}$ is a linear map D such that $D[X, Y] = [DX, Y] + [X, DY]$. It is easily seen that if D_1 and D_2 are derivations on $\mathfrak{g}$, then so is $[D_1, D_2]$. For $X \in \mathfrak{g}$, ad X is a derivation, called an *inner derivation*. (See exercise 5, §10.)

Theorem 9.17. *All derivations of a semi-simple algebra are inner.*

PROOF. Denote the algebra of derivations on $\mathfrak{g}$ by $\mathfrak{D}(\mathfrak{g})$. The adjoint representation of $\mathfrak{g}$ is a homomorphism of $\mathfrak{g}$ into $\mathfrak{D}(\mathfrak{g})$ whose image is denoted by ad $\mathfrak{g}$. Since $\mathfrak{g}$ is semi-simple the kernel of this homomorphism is trivial; and ad $\mathfrak{g}$ isomorphic to $\mathfrak{g}$, so is also semi-simple. Furthermore, since $[D, \text{ad } X] = \text{ad } DX$, ad $\mathfrak{g}$ is an ideal in $\mathfrak{D}(\mathfrak{g})$. Its orthogonal complement (relative to the Killing form in $\mathfrak{D}(\mathfrak{g})$) is also an ideal, and $\mathfrak{a} = (\text{ad } \mathfrak{g}) \cap (\text{ad } \mathfrak{g})^\perp$ is therefore an ideal in ad $\mathfrak{g}$. Since the Killing form vanishes identically on $\mathfrak{a}$, $\mathfrak{a}$ is solvable by Theorem 9.15. Since ad $\mathfrak{g}$ is semi-simple, $\mathfrak{a} = 0$.

Now for any $D \in (\text{ad } \mathfrak{g})^\perp$, ad $DX = [D, \text{ad } X] \in \mathfrak{a}$, hence ad $DX = 0$ for all $D \in (\text{ad } \mathfrak{g})^\perp$. Since the kernel of the adjoint representation is trivial, this means $DX = 0$ for any such D, and so $(\text{ad } \mathfrak{g})^\perp$ is empty. ☐

For further reading in the structure of Lie algebras, see Jacobson or Samelson.

CHAPTER 10

Structure of Semi-Simple Lie Algebras

33. Weyl–Chevalley Normal Form

The semi-simple algebras and their representations play a central role in applications to physical examples. For example, $so(3)$, $su(2)$, $so(3, 1)$ and $su(3)$ are all (real) semi-simple algebras. The complex semi-simple algebras were originally classified by Killing and Cartan in his thesis (1894); and later (1914) Cartan classified the real ones. We discuss these matters in some detail in this chapter; but before launching into an abstract algebraic development, we ease into the subject with two simple examples.

We begin with the real algebra $su(2)$, whose commutation relations are $[L_i, L_j] = \varepsilon_{ijk} L_k$. Let us choose one of these, say L_3, and consider ad L_3. Its matrix is given by

$$\begin{pmatrix} 0 & -1 & 0 \\ 1 & 0 & 0 \\ 0 & 0 & 0 \end{pmatrix}$$

and so the eigenvalues of ad L_3 are 0, $\pm i$. Clearly we cannot diagonalize it over the reals, so we complexify $su(2)$, obtaining $sl(2, C)$. We now choose as a basis for $sl(2, C)$ the operators $J_3 = iL_3$, $J_\pm = \pm L_1 + iL_2$. The commutation relations of these operators are

$$[J_3, J_\pm] = \pm J_\pm, \qquad [J_+, J_-] = 2J_3.$$

In this basis ad J_3 is a diagonal operator with eigenvalues 0, ± 1. Its eigenvectors are J_-, J_3, and J_+, and $sl(2, C)$ is decomposed into a direct sum of invariant subspaces of ad J_3.

We represent the situation in Fig. 10.1. Note that ad $J_+(J_-) = 2J_3$,

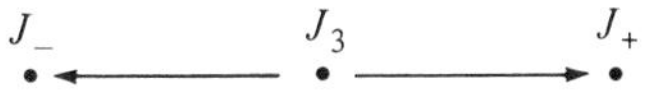

Figure 10.1. The simple algebra A_1.

$\operatorname{ad} J_+(J_3) = -J_+$, and $\operatorname{ad} J_+(J_+) = 0$. Thus $\operatorname{ad} J_+$ is nilpotent and acts by shifting $\{J_-, J_3, J_+\}$ to the right until it finally annihilates the vector at the last stage. Similarly $\operatorname{ad} J_-$ shifts everything to the left. The operator $\operatorname{ad} J_3$ is semi-simple and $\operatorname{ad} J_+$ are nilpotent. Note, too, that we have found a basis for $sl(2, C)$ in which the commutation relations are real integers.

The diagram in Fig. 10.1 is the vector figure for A_1 and is the first in the series A_l of classical Lie algebras. Now let us pass to the more complicated situation A_2.

The algebra A_2 is the complex algebra $sl(3, C)$. Denote by $\mathfrak{g}_0$ the subalgebra of all diagonal matrices of trace zero. It is clear that $\mathfrak{g}_0$ is a two dimensional abelian subalgebra of $sl(3, C)$. We shall see that the family $\{\operatorname{ad} H | H \in \mathfrak{g}_0\}$ is a commuting family of semi-simple operators. The operators $\operatorname{ad} H$ are simultaneously diagonalizeable and the algebra $sl(3, C)$ is decomposable into a direct sum of their one-dimensional invariant subspaces. The algebra $\mathfrak{g}_0$ is a *Cartan subalgebra* of $sl(3, C)$.

The eigenvalues of $\operatorname{ad} H$ are given by the roots of the characteristic polynomial

$$\det(\alpha\mathbb{1} - \operatorname{ad} H) = 0.$$

The solutions of this equation are in fact functionals of H and therefore are to be regarded as functionals on the Cartan subalgebra $\mathfrak{g}_0$. They are called *roots* of the algebra, and the corresponding eigenvectors are called *root vectors*. Thus, if α is a root and E_α the root vector, we have

$$[H, E_\alpha] = \alpha(H)E_\alpha$$

for all $H \in \mathfrak{g}_0$. It is immediate from this that the roots are linear functionals on $\mathfrak{g}_0$.

We choose a basis for $sl(3, C)$ as follows. Let E_{ij} be the Weyl basis for $gl(3, C)$; that is, E_{ij} has a 1 in the ijth slot and zeros elsewhere. Let

$$E_\alpha = E_{12} \qquad E_\beta = E_{23} \qquad E_\gamma = E_{13}$$

$$E_{-\alpha} = E_{21} \qquad E_{-\beta} = E_{32} \qquad E_{-\gamma} = E_{31}$$

$$H_{12} = \begin{pmatrix} 1 & 0 & 0 \\ 0 & -1 & 0 \\ 0 & 0 & 0 \end{pmatrix} \qquad H_{23} = \begin{pmatrix} 0 & 0 & 0 \\ 0 & 1 & 0 \\ 0 & 0 & -1 \end{pmatrix}.$$

If H is any diagonal matrix with trace zero, say $H = \operatorname{diag}(\lambda_1, \lambda_2, \lambda_3)$, we easily find that

$$[H, E_{ij}] = (\lambda_i - \lambda_j)E_{ij}.$$

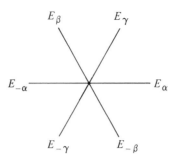

Figure 10.2. Root figure for A_2.

We define the functionals ω_i on the Cartan subalgebra to be $\omega_i(H) = \lambda_i$. Then $\omega_{ij} = \omega_i - \omega_j$ are the roots of the subalgebra with root vectors E_{ij}.

We define an inner product on the dual space $\mathfrak{g}_0'$ as follows. If α is a linear functional on $\mathfrak{g}_0$ there is a unique diagonal matrix H_α such that $\alpha(H) = \operatorname{tr} HH_\alpha$. In fact, if $\alpha = \omega_i$ then H_α is the diagonal matrix with a 1 in the iith entry and zeros elsewhere. Thus $\omega_{12}(H) = \operatorname{Tr} HH_{12}$ and $\omega_{23}(H) = \operatorname{Tr} HH_{23}$, etc. The inner product on $\mathfrak{g}_0'$ is then given by $\langle \alpha, \beta \rangle = \operatorname{Tr} H_\alpha H_\beta$.

We let $\alpha = \omega_{12}$, $\beta = \omega_{23}$, and $\gamma = \omega_{13}$. It is immediate that $\alpha + \beta = \gamma$. Moreover, $\langle \alpha, \alpha \rangle = \langle \beta, \beta \rangle = \langle \gamma, \gamma \rangle = 2$ and $\langle \alpha, \beta \rangle = -1$. The roots $\pm \alpha$, $\pm \beta$, $\pm \gamma$ may thus be depicted as in Fig. 10.2. As in the case of the algebra A_1 we associate the elements $E_{\pm\alpha}, E_{\pm\beta}, E_{\pm\gamma}$ with the roots $\pm \alpha, \pm \beta, \pm \gamma$. Then ad E_α shifts everything in the direction α; that is, ad $E_\alpha(E_\beta) = E_\gamma$ and ad $E_\alpha(E_\gamma) = 0$. Similarly, ad E_γ shifts everything in the direction γ. The operators ad H_1 and ad H_2 are semi-simple, and the operators ad E_α, ad E_β, etc. are nilpotent.

As before, we have found a basis for the algebra $sl(3, C)$ relative to which all the structure constants are real integers. These two examples are proto-typical of the structure of all the semi-simple algebras.

A Cartan subalgebra of a semi-simple algebra $\mathfrak{g}$ is a maximal abelian subalgebra $\mathfrak{g}_0$ such that ad H is semi-simple for all $H \in \mathfrak{g}_0$. Since the operators $\{\text{ad } H | H \in h\}$ are a commuting family of semi-simple operators, there is a basis for $\mathfrak{g}$ in which these operators are simultaneously diagonalizeable. The rank of $\mathfrak{g}$ is the dimension of its Cartan subalgebra. The *roots* of $\mathfrak{g}$ are the functionals α on $\mathfrak{g}_0$ such that $[H, E_\alpha] = \alpha(H)E_\alpha$ for some vector E_α in $\mathfrak{g}$; and the vectors E_α are called the *root vectors*. It turns out that each of the invariant subspaces is one-dimensional. Moreover, $\mathfrak{g}$ is decomposed into a direct sum of these root spaces:

$$\mathfrak{g} = \mathfrak{g}_0 + \bigoplus_\alpha \mathfrak{g}_\alpha$$

where $\mathfrak{g}_0$ is the Cartan subalgebra. The root spaces $\mathfrak{g}_\alpha$ satisfy $[\mathfrak{g}_\alpha, \mathfrak{g}_\beta] = \mathfrak{g}_{\alpha+\beta}$ where $\mathfrak{g}_{\alpha+\beta} = 0$ if $\alpha + \beta$ is not a root.

The existence of a Cartan subalgebra will be proved in §34. The basic structure of a semi-simple algebra is given by the following:

Theorem 10.1 (Weyl–Chevalley Normal Form). *Let $\mathfrak{g}$ be a complex semi-simple algebra and let $\mathfrak{g}_0$ be its Cartan subalgebra. The algebra $\mathfrak{g}$ decomposes into a direct sum of $\mathfrak{g}_0$ plus its one-dimensional root spaces $\mathfrak{g}_\alpha$ such that*

(i) $[H, E_\alpha] = \alpha(H)E_\alpha$ *for* $H \in \mathfrak{g}_0$

(ii) $[E_\alpha, E_{-\alpha}] \in \mathfrak{g}_0$

(iii) $[E_\alpha, E_\beta] = N_{\alpha\beta}E_{\alpha+\beta}$ *where* $N_{\alpha\beta} = 0$ *unless* $\alpha + \beta$ *is a root.*

There exists a basis for $\mathfrak{g}$ in which all the structure constants are real integers and $N_{\alpha\beta} = -N_{(-\alpha)(-\beta)}$.

The observation that a basis exists in which the structure constants are integers is due to Chevalley. This theorem will be proved (except for the integer normalization) in §34.

EXERCISES

1. The $(l + 1) \times (l + 1)$ matrices of trace zero form the Lie algebra $sl(l + 1, C)$, denoted by A_l in Cartan's notation. Take as a basis the matrices E_{ij} for $i \ne j$ and $H_i = E_{ii} - E_{(i+1)(i+1)}$ where $i = 1, \ldots, l$. Show

   (i) $\mathfrak{g}_0 = \{H_1, \ldots, H_l\}$ is an abelian algebra and each ad H_i is semi-simple.
   (ii) If $H = \text{diag}(\lambda_1, \ldots, \lambda_{l+1})$ show that ad $H(E_{ij}) = (\lambda_i - \lambda_j)E_{ij}$.

   Show the roots of A_l are given by the functionals $\omega_{ij} = \omega_i - \omega_j$, where $\omega_i(H) = \lambda_i$. Show that a basis for the roots is $\omega_{12}, \omega_{23}, \ldots, \omega_{l(l+1)}$. When is $\omega_{ij} + \omega_{mn}$ a root?

   (iii) Show that the Killing form K is given by $K(A, B) = 2(l + 1)\text{Tr } AB$.
   (iv) Let θ_{ij} be the angle between H_i and H_j. Find all θ_{ij}.

$$\left(\cos \theta_{ij} = \frac{K(H_i, H_j)}{\sqrt{K(H_i, H_i)} \sqrt{K(H_j, H_j)}} \right)$$

2. The vector figure (10.3) for the simple algebra B_2 is

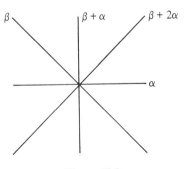

Figure 10.3

Show the root vectors E_β, $E_{\beta+\alpha}$, $E_{\beta+2\alpha}$ form an abelian subalgebra. Why is this not the Cartan subalgebra? (The Cartan subalgebra is two dimensional.) Show that the root vectors E_α, E_β, $E_{\beta+\alpha}$, $E_{\beta+2\alpha}$ from a nilpotent algebra, and that these vectors together with the Cartan subalgebra form a solvable algebra. (*Hint*: these results can be derived entirely from Theorem 10.1.)

34. Cartan Subalgebras, and Root Space Decompositions

We prove here that every semi-simple Lie algebra $\mathfrak{g}$ has a Cartan subalgebra. For $X \in \mathfrak{g}$ we may decompose $\mathfrak{g}$ into a direct sum of invariant subspaces of ad X:

$$\mathfrak{g} = \bigoplus_\lambda \mathfrak{g}_\lambda(X)$$

where

$$\mathfrak{g}_\lambda(X) = \{Y | (\text{ad } X - \lambda)^k Y = 0 \text{ for some } k\}.$$

This is a standard result in linear algebra. We denote by $\mathfrak{g}_0(X)$ the subspace corresponding to $\lambda = 0$ (0 is of course always an eigenvalue of ad X). The *nility* of ad X is the dimension of $\mathfrak{g}_0(X)$. From our examples in §33, we should expect the nility to be smallest for elements of the Cartan subalgebra, since these are all semi-simple while the root vectors are nilpotent. We are therefore led to define a *regular element* to be one whose nility is smallest possible.

Theorem 10.2. *If H is a regular element of a semi-simple Lie algebra $\mathfrak{g}$, then $\mathfrak{g}_0(H)$ is a Cartan subalgebra of $\mathfrak{g}$.*

We shall need the following in our development:

$$[\mathfrak{g}_\lambda(X), \mathfrak{g}_\mu(X)] \subseteq \mathfrak{g}_{\lambda+\mu}(X). \tag{10.1}$$

The proof is outlined in exercise 1 at the end of this section.
We establish Theorem 10.2 in several steps.

Lemma 10.3. *If H is regular then $\mathfrak{g}_0(H)$ is nilpotent.*

PROOF. By Engel's theorem it suffices to prove that every element of $\mathfrak{g}_0(H)$ is ad-nilpotent. Set $H_1 \in \mathfrak{g}_0(H)$ and define $H(z) = zH_1 + (1 - z)H$. These elements all belong to $\mathfrak{g}_0(H)$; and, by (10.1), ad $H(z)$ leaves $\mathfrak{g}_\lambda(X)$ invariant for each λ. For $z = 0$ ad $H(z)$ is a non-singular transformation of each $\mathfrak{g}_\lambda$ for $\lambda \neq 0$. By continuity, ad $H(z)$ is non-singular on each $\mathfrak{g}_\lambda$ for z sufficiently small. Therefore, $\mathfrak{g}_0(H(z)) \cap \mathfrak{g}_\lambda = 0$ for $\lambda \neq 0$, and $\mathfrak{g}_0(H(z)) \subseteq \mathfrak{g}_0(H)$. By the regularity of H, however, we have dim $\mathfrak{g}_0(H(z)) = $ dim $\mathfrak{g}_0(H)$; and therefore $\mathfrak{g}_0(H(z)) = \mathfrak{g}_0(H)$ for small z. It follows that ad $H(z)$ is nilpotent on $\mathfrak{g}_0(H)$ for small z, *viz.*

$$(\text{ad } zH_1 + (1 - z)H)^k(Y) = 0$$

for some k and all small z. Since this is a polynomial in z that vanishes identically for small z, it must vanish for all z; and consequently $(\text{ad } H_1)$ is nilpotent on $\mathfrak{g}_0(H)$. Moreover $\mathfrak{g}_0(H_1) = \mathfrak{g}_0(H)$. □

From now on we let $\mathfrak{g}_0 = \mathfrak{g}_0(H)$ for some regular element H. Since $\mathfrak{g}_0$ is nilpotent there exists a basis for $\mathfrak{g}$ relative to which the operators $\text{ad } H$ are upper triangular. We may write

$$\text{ad } H = \begin{pmatrix} \lambda_1(H) & \cdot & \cdot & \cdot & \cdot \\ & \cdot & \cdot & & \cdot \\ 0 & & & & \lambda_n(H) \end{pmatrix}$$

where $\lambda_i(H)$ are linear functionals, called roots, on the vector space $\mathfrak{g}_0$. If $l = \dim \mathfrak{g}_0$ we may take $\lambda_1 = \cdots = \lambda_l = 0$, corresponding to the fact that $\text{ad } H$ annihilates $\mathfrak{g}_0$ for every H in $\mathfrak{g}_0$, and so has an l dimensional null space.

Lemma 10.4. *For fixed $H \in \mathfrak{g}_0$, $K(X, H) = 0$ for any $X \in \mathfrak{g}_\lambda(H)$, $\lambda \neq 0$. If $\mathfrak{g}$ is semi-simple and if $\lambda_i(H) = 0$ for all the roots λ_i, then $H = 0$. Accordingly, the roots λ_i span the dual space $\mathfrak{g}_0'$.*

PROOF. By (10.1) $\text{ad } X: \mathfrak{g}_\mu \to \mathfrak{g}_{\lambda+\mu}$; therefore $\text{ad } X \text{ ad } H: \mathfrak{g}_\mu \to \mathfrak{g}_{\mu+\lambda}$; $K(X, H) = \text{Tr ad } X \text{ ad } H = 0$; and H is orthogonal to all $X \in \mathfrak{g}_\lambda(H)$ for $\lambda \neq 0$. If all the weights λ_i vanish on H then $\text{ad } H$ is strictly upper triangular. Hence $K(H, H') = 0$ for all $H' \in \mathfrak{g}_0$. This fact combined with the first statement of the lemma and the fact that K is nonsingular if $\mathfrak{g}$ is semi-simple, shows that $H = 0$. The proof of the final statement is left to the reader. □

To prove Theorem 10.2 we must show that if $\mathfrak{g}$ is semi-simple then $\mathfrak{g}_0$ is a maximal abelian sub-algebra and $\text{ad } H$ is semi-simple for all $H \in \mathfrak{g}_0$. Since $\text{ad } H$ is upper triangular for $H \in \mathfrak{g}_0$, $K([H_1, H_2], H) = 0$ for all $H_1, H_2, H \in \mathfrak{g}_0$. By Lemma 10.4, $K([H_1, H_2], W) = 0$ for all $W \in \mathfrak{g}_\lambda([H_1, H_2])$. Therefore, since K is non-degenerate, $[H_1, H_2] = 0$ and $\mathfrak{g}_0$ is abelian. It is maximal by its very definition.

Now let us show that $\text{ad } H$ is semi-simple. We may write $\text{ad } H = S + N$ where S is semi-simple and N is nilpotent. We first show that S is a derivation. In fact, for $X \in \mathfrak{g}_\lambda(H)$ and $Y \in \mathfrak{g}_\mu(H)$ we have $[X, Y] \in \mathfrak{g}_{\lambda+\mu}(H)$, so

$$S[X, Y] = (\lambda(H) + \mu(H))[X, Y] = [SX, Y] + [X, SY].$$

By Theorem 9.17, S is an inner derivation, and $S = \text{ad } W$ for some W. Furthermore, S is a polynomial in $\text{ad } H$, so $\text{ad } W = p(\text{ad } H)$ for some polynomial p. For any $H_1 \in \mathfrak{g}_0$, $\text{ad}[W, H_1] = [S, \text{ad } H_1] = [p(\text{ad } H), H_1] = 0$, since we have shown that $\mathfrak{g}_0$ is abelian. Since $\mathfrak{g}$ is semi-simple its adjoint representation is 1-1, so $[W, H_1] = 0$ for all $H_1 \in \mathfrak{g}_0$; and therefore $W \in \mathfrak{g}_0$ by maximality. Now $\text{ad}(H - W) = N$ is nilpotent so has only zero eigenvalues. Thus $\lambda_i(H - W) = 0$ for all the roots, and by Lemma 10.4, $H - W = 0$. Therefore $N = 0$ and $\text{ad } H$ is semi-simple. □

Given a semi-simple algebra $\mathfrak{g}$ with Cartan subalgebra $\mathfrak{g}_0$, let P denote the set of non-zero roots of $\mathfrak{g}_0$ and for $\alpha \in P$ let

$$\mathfrak{g}_\alpha = \{X \,|\, \text{ad } H(X) = \alpha(H)X\}$$

denotes the root spaces. A vector $X_\alpha \in \mathfrak{g}_\alpha$ is called a root vector. The algebra $\mathfrak{g}$ decomposes into a direct sum of the $\mathfrak{g}_\alpha$:

$$\mathfrak{g} = \mathfrak{g}_0 \oplus \sum_{\alpha \in P} \mathfrak{g}_\alpha.$$

From (10.1) we have

$$[\mathfrak{g}_\alpha, \mathfrak{g}_\beta] \subseteq \mathfrak{g}_{\alpha+\beta}. \tag{10.2}$$

In particular, $[\mathfrak{g}_\alpha, \mathfrak{g}_\beta] = 0$ if $\alpha + \beta$ is not a root. Thus, if $X_\alpha \in \mathfrak{g}_\alpha$, ad X_α maps $\mathfrak{g}_\beta$ to $\mathfrak{g}_{\beta+\alpha}$, $\mathfrak{g}_{\beta+\alpha}$ to $\mathfrak{g}_{\beta+2\alpha}$, and so on. Eventually $\beta + n\alpha$ is not a root, and so ad X_α is nilpotent for any $\alpha \neq 0$.

For example, consider the algebra A_l discussed in exercise 1, §33. The root spaces are the matrices E_{ij} with $i \neq j$ and the roots are $\omega_{ij} = \omega_i - \omega_j$. All the ad E_{ij} are nilpotent. Let ω_{ij} and ω_{mn} be roots; then $\omega_{ij} + \omega_{mn}$ is a root iff $j = m$ or $i = n$ (or both). For example, if $j = m$ then $\omega_{ij} + \omega_{jn} = \omega_{in}$. On the other hand, $[E_{ij}, E_{mn}] = \delta_{jm}E_{in} - \delta_{ni}E_{mj}$; so $[E_{ij}, E_{mn}] = 0$ unless $j = m$ or $n = i$. If $j = m$ then $[E_{ij}, E_{mn}] = E_{in}$. If $j = m$ and $n = i$ then $\omega_{ij} + \omega_{ji} = 0$, while $[E_{ij}, E_{mn}] = E_{ii} - E_{jj}$.

We summarize some further properties of the root spaces in the following lemma. The Killing form is denoted by K.

Lemma 10.5. (i) $\mathfrak{g}_\alpha \perp \mathfrak{g}_\beta$ if $\alpha + \beta \neq 0$.

(ii) *The restriction of K to $\mathfrak{g}_0$ is non-degenerate, and for each root α there exists a unique $H_\alpha \in \mathfrak{g}_0$ such that $\alpha(H) = K(H, H_\alpha)$.*

(iii) *If $\alpha \in P$ then so does $-\alpha$; and for $X \in \mathfrak{g}_\alpha$ and $Y \in \mathfrak{g}_{-\alpha}$, $[X, Y] = K(X, Y)H_\alpha$.*

(iv) $\alpha(H_\alpha) = K(H_\alpha, H_\alpha) \neq 0$.

PROOF. (i) This follows from the identity

$$(\alpha(H) + \beta(H))K(X_\alpha, X_\beta) = K([H, X_\alpha], X_\beta) + K(X_\alpha, [H, X_\beta]) = 0.$$

(ii) Given $H \in \mathfrak{g}_0$, $K(H, X_\alpha) = 0$ for all $\alpha \neq 0$ by (i). Therefore if $K(H, H') = 0$ for all $H' \in \mathfrak{g}_0$ as well, then $K(H, W) = 0$ for all $W \in \mathfrak{g}$. This implies $H = 0$ by the non-degeneracy of K. Thus K restricted to $\mathfrak{g}_0$ is non-degenerate, and K effects the duality between $\mathfrak{g}_0$ and its dual. Given any $\alpha \in P$ there is an $H_\alpha \in \mathfrak{g}_0$ such that $\alpha(H) = K(H, H_\alpha)$ for all $H \in \mathfrak{g}_0$.

(iii) If $-\alpha$ is not a root then $\alpha + \beta$ is never zero for any root β, so $K(X_\alpha, Y_\beta) = 0$ for all Y_β by (i); hence $X_\alpha = 0$.

If $X \in \mathfrak{g}_\alpha$ and $Y \in \mathfrak{g}_{-\alpha}$ then $[X, Y] \in \mathfrak{g}_0$ by (10.2). Moreover,

$$K([X, Y], H) = K(X, [Y, H]) = K(X, \alpha(H)Y))$$

$$= K(X, Y)\alpha(H) = K(X, Y)K(H, H_\alpha).$$

Therefore $K([X, Y] - K(X, Y)H_\alpha, H) = 0$ for all $H \in \mathfrak{g}_0$. Since K is non-degenerate on H, we have $[X, Y] = K(X, Y)H_\alpha$.

(iv) Let $X_{-\alpha} \in \mathfrak{g}_{-\alpha}$ and suppose $K(X, X_{-\alpha}) = 0$ for all $X \in \mathfrak{g}_0$. Since $K(Y, X_{-\alpha}) = 0$ for all $Y \notin \mathfrak{g}_\alpha$ this would imply $X_{-\alpha} = 0$. Therefore there exists a vector $X_\alpha \in \mathfrak{g}_\alpha$ such that $K(X_\alpha, X_{-\alpha}) = 1$. By (iii) we have $[X_\alpha, X_{-\alpha}] = H_\alpha$. $\square$

Now consider the string $\mathfrak{g}_{\beta\alpha} = \bigoplus_j \mathfrak{g}_{\beta+j\alpha}$. By (10.2) these subspaces are invariant under ad X_α and ad $X_{-\alpha}$. Since ad H_α also leaves this string invariant, $\operatorname{Tr} \operatorname{ad} H_\alpha = \operatorname{Tr} \operatorname{ad}[X_\alpha, X_{-\alpha}] = \operatorname{Tr}[\operatorname{ad} X_\alpha, \operatorname{ad} X_{-\alpha}] = 0$, where the trace is computed relative to the subspace $\mathfrak{g}_{\beta\alpha}$. On the other hand,

$$\operatorname{Tr} \operatorname{ad} H_\alpha = \sum_j (\beta + j\alpha)(H_\alpha)\dim \mathfrak{g}_{\beta+j\alpha},$$

hence

$$\beta(H_\alpha)\dim \mathfrak{g}_{\beta\alpha} + \alpha(H_\alpha)\sum_j j \dim \mathfrak{g}_{\beta+j\alpha} = 0.$$

If $\alpha(H_\alpha) = 0$ then $\beta(H_\alpha) = 0$ for all $\beta \in P$; and, by Lemma 10.4, it would follow that $H_\alpha = 0$. Therefore $\alpha(H_\alpha) \neq 0$.

Since $K(H_\alpha, H_\alpha) \neq 0$ we may define

$$T_\alpha = \frac{2}{K(H_\alpha, H_\alpha)} H_\alpha \qquad \alpha \in P.$$

Note that $\alpha(T_\alpha) = 2$. The numbers $a_{\beta\alpha}$ defined by

$$a_{\beta\alpha} = \beta(T_\alpha) = \frac{2K(H_\beta, H_\alpha)}{K(H_\alpha, H_\alpha)}$$

are integers, as we shall see, called the *Cartan integers*. From the proof of Lemma 10.5 (iv) we may choose $X_\alpha \in \mathfrak{g}_\alpha$ and $X_{-\alpha} \in \mathfrak{g}_{-\alpha}$ so that $K(X_\alpha, X_{-\alpha}) \neq 0$; we choose these vectors so that

$$K(X_\alpha, X_{-\alpha}) = \frac{2}{K(H_\alpha, H_\alpha)}.$$

Having done this, we see that $\{X_\alpha, X_{-\alpha}, T_\alpha\}$ forms a real Lie algebra with commutation relations

$$[X_\alpha, X_{-\alpha}] = T_\alpha, \qquad [T_\alpha, X_{\pm\alpha}] = \pm 2X_{\pm\alpha}.$$

This is the algebra $sl(2, R)$, which arose in Chapter 4 in the discussion of the quantization of angular momentum. We need

Lemma 10.6. *Let* $\{T, X_+, X_-\}$ *satisfy the* $sl(2, R)$ *commutation relations and let the algebra act irreducibly on a vector space* V. *Then there is a positive integer* r *such that* $V = sp[v_0, v_1, \ldots, v_r]$ *where* $Tv_j = (r - 2j)v_j$, *and* $X_- v_j = v_{j+1}$.

PROOF. Since V is finite dimensional the operator T has an eigenvalue, say $Tv = \lambda v$. The commutation relations show that $TX_+ v = (\lambda \pm 2)X_+ v$; thus X_+ act as shift operators on the eigenvectors of T.

Since the representation is finite dimensional there exists a highest vector v_0 such that $Tv_0 = \lambda v_0$ and $X_+ v_0 = 0$. Define $v_j = (X_-)^j v_0$. Then $Tv_j = (\lambda - 2j)v_j$; and there is a lowest vector v_r such that $X_- v_r = 0$. One (i.e. you) may prove by induction that $X_+ v_j = j(\lambda - j + 1)v_{j-1}$. Taking $j = r + 1$ we get $v_{r+1} = 0$ so $X_+ v_{r+1} = 0 = (r + 1)(\lambda - r)v_r$. This shows that $\lambda = r$. It is clear that the algebra acts irreducibly on $\{v_0, \ldots, v_r\}$, so these vectors span V. $\qquad\square$

We use the fact that $\{T_\alpha, X_\alpha, X_{-\alpha}\}$ acts on the strings $\mathfrak{g}_{\beta\alpha} = \bigoplus_j \mathfrak{g}_{\beta+j\alpha}$ via the adjoint representation to gain more information about these subspaces. In particular, the following lemma shows that $\{T_\alpha, X_\alpha, X_{-\alpha}\}$ acts irreducibly on $\mathfrak{g}_{\beta\alpha}$.

Lemma 10.7. (i) *For any roots α, $\beta \in P$ the string $\{\beta + n\alpha\}$ is an uninterrupted sequence of roots for $p \le n \le q$, with $p \le 0$ and $q \ge 0$. Moreover, $a_{\beta\alpha} = -(p + q)$, and there are no other roots of the form $\beta + \mu\alpha$.*

(ii) *For $\alpha \in P$, $\dim \mathfrak{g}_\alpha = 1$ and the only other roots proportional to α are 0 and $-\alpha$.*

(iii) $[\mathfrak{g}_\alpha, \mathfrak{g}_\beta] = \mathfrak{g}_{\alpha+\beta}$.

(iv) *For any two roots α and β, let $\varepsilon = \operatorname{sgn} a_{\beta\alpha}$. Then $\beta - \varepsilon\alpha$, $\beta - \varepsilon 2\alpha$, $\ldots$, $\beta - \alpha_{\beta\alpha}$ are all roots.*

PROOF. (i) The space $\mathfrak{g}_{\beta\alpha} = \bigoplus_n \mathfrak{g}_{\beta+n\alpha}$ is invariant under the adjoint action of $\{X_\alpha, X_{-\alpha}, T_\alpha\}$. We are going to see, in the course of this lemma, that this action is irreducible; but we don't know that *a priori*. The space $\mathfrak{g}_{\beta\alpha}$ consists of the vectors $(\operatorname{ad} X_\alpha)^j X_\beta$ and $(\operatorname{ad} X_{-\alpha})^k X_\beta$ for some integers j and k. Since $(\operatorname{ad} T_\alpha)(X_\beta) = \beta(T_\alpha)X_\beta$, the eigenvalues of $\operatorname{ad} T_\alpha$ on $\mathfrak{g}_{\beta\alpha}$ are $\beta(T_\alpha) + n\alpha(T_\alpha) = a_{\beta\alpha} + 2n$ for some set of integers n, $p \le n \le q$, where $p \le 0$ and $q \ge 0$. By Lemma 10.6 the largest and smallest eigenvalues are $\pm r$; so p and q are determined by

$$a_{\beta\alpha} + 2p = -r, \qquad a_{\beta\alpha} + 2q = r.$$

It follows that $a_{\beta\alpha} = -(p + q)$; thus the Cartan integers are in fact integers.

We have an uninterrupted string of roots $\beta + n\alpha$ for $p \le n \le q$. There can be no other roots of the form $\beta + m\alpha$, $p' \le m \le q'$, with $q < p'$ or $q' < p$; for in that case we would have a second representation of $\{X_\alpha, X_{-\alpha}, T_\alpha\}$, and $\beta(T_\alpha)$ would be equal also to $-(p' + q')$, which is impossible. The only other possibility is that some of the roots $\beta + n\alpha$ are multiple roots, but this will be excluded in (ii) below. Hence $\{X_\alpha, X_{-\alpha}, T_\alpha\}$ acts irreducibly.

(ii) Let S_α be the subspace spanned by $X_{-\alpha}$, T_α, and $\mathfrak{g}_{n\alpha}$ for $n = 1, 2, \ldots$. Then S_α is invariant under $\operatorname{ad} X_{\pm\alpha}$ and $\operatorname{ad} T_\alpha$. Now $\operatorname{Tr} \operatorname{ad} T_\alpha = \operatorname{Tr} \operatorname{ad}[X_\alpha, X_{-\alpha}] = \operatorname{Tr}[\operatorname{ad} X_\alpha, \operatorname{ad} X_{-\alpha}] = 0$; while $\operatorname{Tr} \operatorname{ad} T_\alpha = -\alpha(T_\alpha) + \sum_{n\ge 1} n\alpha(T_\alpha)\dim \mathfrak{g}_{n\alpha}$. Therefore $\sum_{n\ge 1} n \dim \mathfrak{g}_{n\alpha} = 1$; and we must have $\dim \mathfrak{g}_\alpha = 1$, and $\dim \mathfrak{g}_{n\alpha} = 0$ for

$n \geq 1$. Therefore no positive integral multiple of α can be a root. Moreover, α/n cannot be a root, for this same argument would imply that $\alpha = n(\alpha/n)$ could not be a root.

If $\beta = c\alpha$ is a root for some complex number c, we know from (i) that $\beta(T_\alpha) = c\alpha(T_\alpha) = 2c$ is an integer. It suffices to consider the case $c > 0$ (for negative c apply the argument to the root $-\alpha$). By the previous paragraph we know c cannot be a positive integer greater than one. The remaining possibility is that $c = r/2$ for some integer r. We leave it as an exercise to show that in this case $((r/2) - j)\alpha$ are all roots, for $j = 0, 1, \ldots, r/2$. (Consider $\mathrm{ad}\, T_\alpha$, $\mathrm{ad}\, X_{\pm\alpha}$ acting on the space $\bigoplus_n \mathfrak{g}_{\beta+n\alpha}$.) Thus if $r\alpha/2$ is a root, then so is $\alpha/2$; but this possibility was excluded in the previous paragraph.

(iii) Since $\dim \mathfrak{g}_\alpha = 1$, the adjoint representation of $\{X_\alpha, X_{-\alpha}, T_\alpha\}$ acts irreducibly on the subspace $\bigoplus_{n=-p}^{q} \mathfrak{g}_{\beta+n\alpha}$. Thus $\mathrm{ad}\, X_\alpha$ can annihilate X_β only if β is at the top of the α string containing β. In this case $q = 0$ and $\beta + \alpha$ is not a root. Thus $[\mathfrak{g}_\alpha, \mathfrak{g}_\beta] = \mathfrak{g}_{\alpha+\beta}$ in all cases. ($\mathfrak{g}_{\alpha+\beta} = 0$ iff $\alpha + \beta$ is not a root.)

(iv) It suffices to show that $\beta - a^\alpha_{\beta\alpha}$ is a root. Since $p \leq 0 \leq q$, $p \leq p + q \leq q$ and so $\beta - a^\alpha_{\beta\alpha} = \beta + (p + q)\alpha$ is a root. $\qquad\square$

We have now established all but the last statement of the Weyl–Chevalley normal form, namely that there exists a basis for $\mathfrak{g}$ in which all the structure constants are real integers. We have shown there exist root vectors E_α and vectors T_α in $\mathfrak{g}_0$ such that

$$[E_\alpha, E_{-\alpha}] = T_\alpha, \qquad [T_\alpha, E_\beta] = a_{\beta\alpha} E_\beta, \qquad [E_\alpha, E_\beta] = N_{\alpha\beta} E_{\alpha+\beta} \qquad (10.3)$$

where $N_{\alpha\beta} = 0$ if $\alpha + \beta$ is not a root. The first two sets of commutation relations are preserved if each E_α is rescaled by a factor C_α, provided that $C_\alpha C_{-\alpha} = 1$. If this is done then $N_{\alpha\beta}$ is rescaled to $C_{\alpha+\beta}$. A scaling can be chosen in which $N_{\alpha\beta} = \pm(-p + 1)$, where $p \leq 0$ is the smallest integer such that $\beta + p\alpha$ is a root. The signs are a little tricky to determine. The procedure is described in Samelson, pp. 51–55. We omit it here since later, in §36, we construct Weyl–Chevalley bases for the classical Lie algebras.

EXERCISES

1. Derive the identity

$$(\mathrm{ad}\, X - (\lambda + \mu)\mathbb{1})^n [A, B] = \sum_{j=0}^{n} \binom{n}{j} [(\mathrm{ad}\, X - \lambda)^j A, (\mathrm{ad}\, X - \mu)^{n-j} B]$$

   and use the result to prove (10.1).

2. Prove that

$$K(H_1, H_2) = \sum_{\alpha \in p} \alpha(H_1)\alpha(H_2) \qquad (10.4)$$

   for $H_1, H_2 \in \mathfrak{g}_0$. Show this bilinear form is positive definite on $\mathfrak{g}_0$.

3. Show that the subspace spanned by T_α, $\alpha \in P$, is a real Lie algebra with a positive definite bilinear form given by (10.4).

4. Let $\mathfrak{g}_0$ be the real vector space spanned by the T_α and let the inner product on $\mathfrak{g}_0$ be denoted by $\langle \alpha, \beta \rangle = K(H_\alpha, H_\beta)$. Show that the reflection in the hyperplane $W_\alpha = \{\beta | \langle \alpha, \beta \rangle = 0\}$ is given by $S_\alpha(\lambda) = \lambda - (2\langle \lambda, \alpha \rangle / \langle \alpha, \alpha \rangle)\alpha = \lambda - \lambda(T_\alpha)\alpha = \lambda + (p + q)\alpha$.

35. Root Figures and Dynkin Diagrams

Let $\mathfrak{h}$ be the real subalgebra spanned by the H_α and let $\mathfrak{h}'$ be its (real) dual space. We define an inner product on $\mathfrak{h}'$ by $\langle \alpha, \beta \rangle = K(H_\alpha, H_\beta)$. It is positive definite (exercise 2, §34). It suffices to define it on the roots, since the roots span $\mathfrak{h}'$. With this inner product the mapping $\alpha \to H_\alpha$ is an isomorphism. The Cartan integers are now given by

$$a_{ij} = \frac{2\langle \alpha_i, \alpha_j \rangle}{\langle \alpha_j, \alpha_j \rangle}$$

where $\{\alpha_j\}$ are the roots.

The collection of roots on a semi-simple algebra $\mathfrak{g}$, sitting in the space $\mathfrak{h}'$, is called the *root diagram*. We are going to see that there are very strong constraints placed on the geometry of the root diagram, and it is these constraints that permit the complete classification of all the complex semi-simple algebras. Recall that P denotes the non-zero roots.

For $\alpha \in P$ let S_α be the reflection in the hyperplane orthogonal to α. S_α is illustrated in Fig. 10.4.

$$S_\alpha(\lambda) = \lambda - \frac{2\langle \lambda, \alpha \rangle \alpha}{\langle \alpha, \alpha \rangle}. \tag{10.5}$$

We summarize the properties of the root diagram in the following theorem.

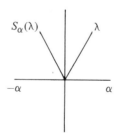

Figure 10.4

Theorem 10.8. (i) *For* $\alpha, \beta \in P$,

$$\beta(T_\alpha) = a_{\beta\alpha} = \frac{2\langle \beta, \alpha \rangle}{\langle \alpha, \alpha \rangle}$$

is an integer.

(ii) *P is invariant under all reflections* S_α, $\alpha \in P$.

(iii) *If* $\alpha \in P$ *and* $r\alpha \in P$ *then* $r = \pm 1$.

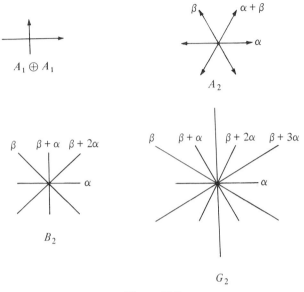

Figure 10.5

Property (ii) is a restatement of Lemma 10.7, (iv): if α and β are roots then $S_\alpha(\beta) = \beta - a_{\beta\alpha}\alpha$ is also a root.

The group of isometries generated by the reflections S_α is called the *Weyl group*. For example, the Weyl group of the vector figure A_2 is the permutation group S_3. A vector figure is simple if it is not the union of two disjoint orthogonal vector figures. For example, the vector figure for $A_1 \oplus A_1$ is not simple (see Fig. 10.5 above). ($A_1 \oplus A_1$ is the vector figure for the complexification of $so(4)$, for example, since $so(4) \cong so(3) \oplus so(3)$.) The rank of a vector figure is the dimension of the subspace spanned by the vectors; we shall always assume this to be the rank of the Cartan subalgebra.

Every semi-simple Lie algebra can be reconstructed from its vector figure; this is the content of Theorem 10.1. All vector figures may be classified and to each vector figure there corresponds a Lie algebra.

The only vector figure of rank 1 is $A_1 = \{-\alpha, 0, \alpha\}$. Let us find all the vector figures of rank 2. Choose α and β independent in P. Then $a_{\alpha\beta}a_{\beta\alpha} = 4\cos^2\theta$ where θ is the angle between α and β. Since $a_{\alpha\beta}$ are integers their product can only be 0, 1, 2, 3, or 4; and 4 can occur only if $\alpha = \pm\beta$. Now assume $a_{\beta\alpha} \le 0$; this can be accomplished by replacing β by $S_\alpha(\beta)$, if necessary. Then either $a_{\beta\alpha} = 0$ and $\alpha \perp \beta$ or one of the numbers $a_{\alpha\beta}$, $a_{\beta\alpha}$ is -1 and the other is -1, -2, or -3. In the second case $a_{\beta\alpha}/a_{\alpha\beta} = |\beta|^2/|\alpha|^2$. Assume that α is the shorter vector (so $a_{\alpha\beta} = -1$), and let θ be the angle between α and β. Then there are four possible vector figures, generated by α and β, corresponding to the four values of $a_{\beta\alpha}$:

	$A_1 \oplus A_1$	A_2	B_2	G_2
$a_{\beta\alpha}$	0	-1	-2	-3
θ	$\pi/2$	$2\pi/3$	$3\pi/4$	$5\pi/6$
$\dfrac{\lvert\beta\rvert}{\lvert\alpha\rvert}$	$?$	1	$\sqrt{2}$	$\sqrt{3}$

Across the top line we have written the four Lie algebras corresponding to the four possible values of $a_{\beta\alpha}$; θ is the angle between α and β. The four vector figures of rank 2 are depicted in Fig. 10.5.

Note that for the choices of α and β in these diagrams $a_{\beta\alpha}$ is negative and $\lvert a_{\beta\alpha}\rvert$ is the number of roots in the α string containing β which lie strictly to the α side of β (these are $\beta + \alpha, \ldots, \beta - a_{\beta\alpha}\alpha$). In no case can there be more than 3; and in fact we are going to see that G_2 is the only algebra in which that can happen.

For semi-simple algebras of higher rank it is no longer possible to represent all the roots in a planar diagram as we have done in Fig. 10.5 for the algebras of rank 2. E.B. Dynkin, in 1944, devised a scheme for representing, and indeed classifying, all the semi-simple algebras by means of planar graphs. This was done by introducing a suitable notion of fundamental roots. We now explain this procedure.

We define an ordering on the non-zero roots as follows. Given an element $H_0 \in \mathfrak{h}$ we define the positive roots to be the set $P^+ = \{\gamma \mid \gamma(H_0) > 0, \gamma \in P\}$. For example, in the case A_2 we take H_0 to be $H_{\alpha+\beta}$. Then $\gamma(H_{\alpha+\beta}) = \langle \gamma, \alpha + \beta\rangle > 0$ iff $\gamma = \alpha, \beta$, or $\alpha + \beta$; thus $P^+ = \{\alpha, \beta, \alpha + \beta\}$. The positive roots of B_2 and G_2 are defined in the same way.

A root $\alpha \in P^+$ is *simple*, or *fundamental*, if it is not the sum of two other roots in P^+. For example, the roots α and β in the root diagrams of Fig. 10.5 are simple.

Given an ordering of the roots, let $F = \{\alpha_1 \ldots \alpha_l\}$ be the set of positive simple roots. Then

Lemma 10.9. *The elements of F form a basis for the real vector space $\mathfrak{h}'$. If $\alpha_i \neq \alpha_j$ belong to F then $\langle \alpha_i, \alpha_j\rangle \leq 0$. Every vector in P^+ is a linear combination of vectors in F with positive integer coefficients.*

PROOF. If $\langle \alpha_i, \alpha_j\rangle > 0$ then $\alpha_i - \alpha_j$ and $\alpha_j - \alpha_i$ are both roots by (iv), Lemma 10.7. One of them, say $\alpha_i - \alpha_j$, is positive; but then $\alpha_i = \alpha_j + (\alpha_i - \alpha_j)$ is not simple. So $\langle \alpha_i, \alpha_j\rangle \leq 0$.

If $\alpha \in P^+$ is not simple it is the sum of two vectors α_1 and α_2 in P^+. If these are not simple they in turn can be reduced. Now $\alpha(H) = \alpha_1(H) + \alpha_2(H)$; and since α_1 and α_2 are both positive, $\alpha_1(H) < \alpha(H)$ and $\alpha_2(H) < \alpha(H)$. Therefore $\alpha(H)$ is decreased with each reduction, and the process must eventually terminate. $\qquad\square$

$$A_1 \oplus A_1 \qquad A_2 \qquad B_2 \qquad G_2$$

Figure 10.6

This shows that the fundamental roots span $\mathfrak{h}'$. To show they are independent, suppose $\sum_i x_i \alpha_i = 0$; and write this as $\sum p_i \alpha_i + \sum n'_j \alpha_j = 0$ where $p_i \geq 0$, and $n'_j \leq 0$. We thus have $\sum p_i \alpha_i = \sum n_j \alpha_j$ where p_i and $n_j (= -n'_j)$ are nonnegative. For $\alpha = \sum p_i \alpha_i$ we have $\langle \alpha, \alpha \rangle = \sum p_i n_j \langle \alpha_i, \alpha_j \rangle \leq 0$; hence $\alpha = 0$. Now for H_0 we have $\alpha(H_0) = \sum p_i \alpha_i(H_0)$, hence all the p_i must vanish. Similarly, all the n_j must vanish.

Now let $F = \{\alpha_1 \ldots \alpha_l\}$ be the fundamental system of roots which generates a vector figure. The Cartan *matrix* with entries

$$a_{ij} = \frac{2\langle \alpha_i, \alpha_j \rangle}{\langle \alpha_j, \alpha_j \rangle}$$

is a matrix of non-positive integers by the above lemma. The set of all fundamental systems can be classified by graphs, called Dynkin diagrams, and to each such diagram corresponds an equivalence class of isomorphic Lie algebras.

If α_i and α_j are simple roots then $a_{ij} a_{ji} = 4 \cos^2 \theta$, where θ is the angle between them. Since the a_{ij} are non-positive integers, the only possible values of $a_{ij} a_{ji}$ are 0, 1, 2, 3, and 4. As before, 4 is excluded since it corresponds to an angle of 0 or π; and α_i and α_j are linearly independent. The subgraph for the pair of roots α_i, α_j consists of two vertices with $a_{ij} a_{ji}$ edges joining them. The four possibilities are shown in Fig. 10.6 corresponding to the four algebras of rank 2 in Fig. 10.5.

The Dynkin diagram of a semi-simple Lie algebra is constructed as follows.

(1) Every root α_i is represented by a vertex.
(2) Every pair of vertices α_i, α_j is joined by $a_{ij} a_{ji}$ edges.
(3) In the case of two or more edges an arrow is drawn in the direction of the longer root.

We may restrict ourselves to connected diagrams. We leave it as an exercise to prove the following theorem.

Theorem 10.10. *Suppose that the root diagram of a Lie algebra $\mathfrak{g}$ decomposes into an orthogonal direct sum $P = P_1 \cup P_2$, where $\langle \alpha, \beta \rangle = 0$ for $\alpha \in P_1$ and $\beta \in P_2$. Then $\mathfrak{g} = \mathfrak{g}_1 \oplus \mathfrak{g}_2$ where $[\mathfrak{g}_1, \mathfrak{g}_2] = 0$ and $\mathfrak{g}_1 \perp \mathfrak{g}_2$ with respect to the Killing form. Accordingly the connected components of the Dynkin diagram correspond to simple algebras; and the union of the connected components corresponds to a direct sum decomposition of $\mathfrak{g}$ into orthogonal ideals.*

Let α_i and α_j be simple roots with α_j the longer root. It is not hard to show that the number of edges joining α_i and α_j is one less than the number of roots in the α_i string through α_j. For example, in the case B_2 the α string through β is $\beta, \beta + \alpha, \beta + 2\alpha$, so α and β are joined by two edges. Similarly, for G_2 the α string through β is $\beta, \beta + \alpha, \beta + 2\alpha, \beta + 3\alpha$.

Now let e_i be the normalized roots: $e_i = \alpha_i/\|\alpha_i\|$. Then $\langle e_i, e_j \rangle = \cos \theta_{ij} = -\frac{1}{2}\sqrt{n_{ij}}$ where n_{ij} is the number of edges joining α_i to α_j in the Dynkin diagram. Associated with each diagram is the quadratic form $\langle x, x \rangle$ where $x = \sum_j x_j e_j$. We have

$$\langle x, x \rangle = \sum_j x_j^2 + 2 \sum_{i<j} x_i x_j \langle e_i, e_j \rangle = \sum_j x_j^2 - \sum_{i<j} x_i x_j \sqrt{n_{ij}}.$$

For example, the quadratic form associated with the diagram
· —— · ═══ · —— · is

$$x_1^2 + x_2^2 + x_3^2 + x_4^2 - (x_1 x_2 + \sqrt{2} x_2 x_3 + x_3 x_4).$$

The quadratic form $\langle x, x \rangle$ must be positive definite for all x; and it is this condition alone that determines all the possible Dynkin diagrams. From now on, by Dynkin diagram we shall understand an admissible diagram.

Lemma 10.11. *Dynkin diagrams have no loops, and each vertex meets at most three lines. Moreover, any pair of vertices joined by a single edge can be merged into one, and the resulting diagram is admissible.*

PROOF. Suppose $\{e_1 \ldots e_k\}$ is a loop and that it is a minimal loop (contains no subloops). Putting $x = e_1 + \cdots + e_k$ we have (with $e_{k+1} = e_1$)

$$\langle x, x \rangle = \sum_{j=1}^k \langle e_j, e_j \rangle + 2 \sum_{j=1}^k \langle e_j, e_{j+1} \rangle \le k + 2(-\tfrac{1}{2})k = 0.$$

(Note that $\langle e_j, e_{j+1} \rangle \le -\frac{1}{2}$.) Since $x \ne 0$ (why?) this is impossible. Now let e be any root and let $\{e_j\}$ be the roots joined to it. All the $\langle e_j, e_k \rangle$ vanish since no edges can join them (loops have already been excluded). Put $v = e - \sum_j \langle e, e_j \rangle e_j$; then $\langle v, e_j \rangle = 0$ and

$$1 = \|e\|^2 = \|v\|^2 + \sum_j \langle e, e_j \rangle^2.$$

Therefore $\sum_j \langle e, e_j \rangle^2 < 1$; but since $\langle e, e_j \rangle^2 = n_j/4$; where n_j is the number of edges joining e to e_j, we have $\sum_j n_j < 4$, and, by integrality, $\sum_j n_j \le 3$.

An immediate corollary is that G_2 is the only Dynkin diagram containing a triple edge. The maximal incidence relations at a root are given by the following subdiagrams.

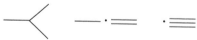

Now let us go to the last statement. We want to prove that if e_1 —— e_2 is part of an admissible diagram then so is the single vertex e_1 joined by all the edges meeting both e_1 and e_2 in the original diagram. An immediate corollary is that none of the following diagrams can occur

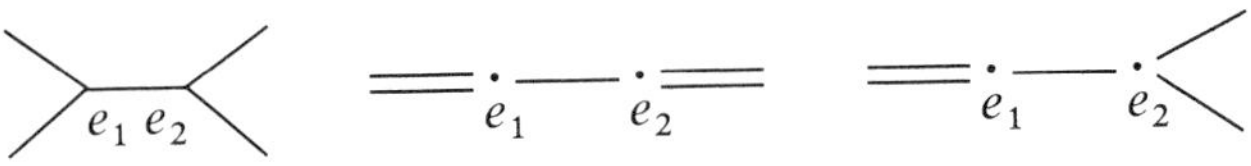

for if the two roots e_1 and e_2 are merged we obtain a graph with four edges meeting a root. Carrying this argument further, no Dynkin diagram can contain the subgraph B_2 more than once; nor can any diagram containing B_2

contain a branch point, i.e. a configuration 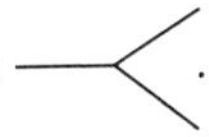 .

The statement is proved by noting that the terms of the quadratic form corresponding to the roots e_1 and e_2 are $x_1^2 + x_2^2 - x_1 x_2 - x_1 p_1 - x_2 p_2$ where p_1 and p_2 are (linear) polynomials in the other variables. If the larger diagram is admissible then the quadratic form is positive definite for all x_1 and x_2. In particular it is still positive definite if we set $x_1 = x_2 = \xi$. Doing this we obtain $\xi^2 - \xi(p_1 + p_2)$, which are the terms of the quadratic form corresponding to the diagram with the vertices merged.   □

Lemma 10.12. *The subdiagram* 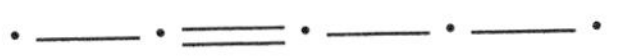 *is impossible.*

PROOF. The quadratic form for this subdiagram is $Q(x) = \sum_{j=1}^5 x_j^2 - (x_1 x_2 + \sqrt{2} x_2 x_3 + x_3 x_4 + x_4 x_5)$ (number the roots from left to right). The critical points of Q are given by the following equations

$$2x_1 = x_2$$
$$2x_2 = x_1 + \sqrt{2} x_3$$
$$2x_3 = \sqrt{2} x_2 + x_4$$
$$2x_4 = x_3 + x_5$$
$$2x_5 = x_4$$

(Note that these can be obtained directly from the incidence relations of the graph: $2x_i = \sum_j \sqrt{n_{ij}} \, x_j$.) A solution is obtained by setting $x_5 = 1$ and solving successively down to x_1. We get $x = (\sqrt{2}, 2\sqrt{2}, 3, 2, 1)$; and for this choice of the x_i we discover that $\langle x, x \rangle = 0$ which is not allowed. Therefore the subdiagram is inadmissible.   □

As an immediate corollary, the only diagrams containing B_2 are

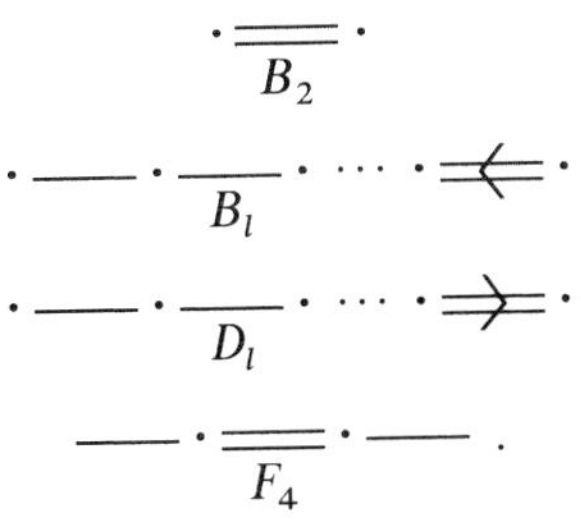

B_2

B_l

D_l

F_4

We have written the notation of the corresponding Lie algebra below the diagram. F_4 is one of the exceptional Lie algebras. One must still show that the corresponding quadratic form is positive definite. In the case of F_4 the form is given by

$$x_1^2 + x_2^2 + x_3^2 + x_4^2 - (x_1 x_2 + \sqrt{2} x_2 x_3 + x_3 x_4).$$

We leave it as an exercise to show this is positive definite.

We have now determined all the graphs with multiple edges. It remains to find the diagrams composed only of single edges. We leave this to exercise 3 below.

A complete list of the possible Dynkin diagrams is given in Theorem 10.13 below. One must also verify, of course, that the quadratic form associated with each diagram is positive definite.

Theorem 10.13. *The complex simple Lie algebras may be classified, up to equivalence, by the following Dynkin diagrams.*

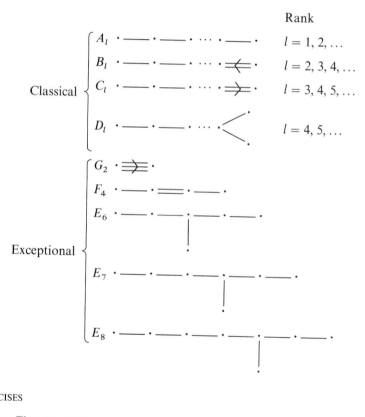

EXERCISES

1. Prove Theorem 10.10.

2. Show the quadratic form for the Dynkin diagram for F_4 is positive definite.

3. Determine which of the following graphs are admissible as follows:

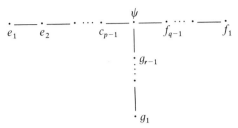

(i) Put $x = \sum_{j=1}^{p-1} j e_j$, $y = \sum_{j=1}^{q-1} j f_j$, $z = \sum_{j=1}^{r-1} j g_j$. Show x, y, z are mutually orthogonal roots and that

$$\|x\|^2 = \frac{p(p-1)}{2}, \qquad \|y\|^2 = \frac{q(q-1)}{2}, \qquad \|z\|^2 = \frac{r(r-1)}{2}.$$

(ii) Let $\theta_1, \theta_2, \theta_3$ be the angles between x, y, z, and ψ. Then $\cos^2 \theta_1 + \cos^2 \theta_2 + \cos^2 \theta_3 < 1$.

(iii) Show $\frac{1}{p} + \frac{1}{q} + \frac{1}{r} > 1$.

(iv) Arrange the ordering so that $p \geq q \geq r$. Then $r = 2$ (if $r = 1$ there is no branch point) and the only solutions are $D_{p+2} = (p, 2, 2)$; $E_6 = (3, 3, 2)$; $E_7 = (4, 3, 2)$; and $E_8 = (5, 3, 2)$.

4. Show that the root vectors corresponding to the positive roots of a semi-simple algebra form a solvable subalgebra.

5. Let α_i and α_j be simple roots with α_j the shorter root. Show that the number of edges joining α_i and α_j in the Dynkin diagram is one less than the number of roots in the α_j string through α_i.

6. Let α_i and α_j be simple (positive) roots. The $\alpha_i - \alpha_j$ and $\alpha_j - \alpha_i$ are not roots.

36. The Classical Algebras

In the previous section we found all the possible Dynkin diagrams for the system of fundamental roots of a simple Lie algebra. There remains the task of showing that an algebra exists for each of the admissible diagrams. We do this here for the classical algebras.

A_l. These are the algebras $sl(l + 1, C)$. We choose a basis for $sl(l + 1, C)$ as follows. Let E_{ij} be the Weyl basis of $gl(n, C)$; that is, E_{ij} is the matrix with a 1 in the ij th slot and zeroes elsewhere. The Cartan subalgebra $\mathfrak{h}$ is given by the diagonal matrices $H = \operatorname{diag}(\lambda_1 \dots \lambda_{l+1})$ with $\operatorname{Tr} H = 0$. It is easily seen that $\operatorname{ad} H(E_{ij}) = [H, E_{ij}] = (\lambda_i - \lambda_j) E_{ij}$. Thus E_{ij} are the root vectors and the roots are the functionals $\alpha_{ij} = \alpha_i - \alpha_j$, where $\alpha_i(H) = \lambda_i$. We define an order on $\mathfrak{h}'$ by choosing any $H_0 \in \mathfrak{h}$ with $\lambda_1 > \lambda_2 > \cdots > \lambda_{l+1}$. Then $\alpha_{ij}(H_0) > 0$ iff $i < j$. The fundamental roots are $\alpha_{12}, \alpha_{23}, \dots, \alpha_{l,l+1}$. To determine the Dynkin diagram, consider the two roots $\alpha_{i,i+1}$ and $\alpha_{j,j+1}$. Adding then we get $\alpha_i - \alpha_{i+1} + \alpha_j - \alpha_{j+1}$; but this is a root only if $j = i + 1$ or $i = j + 1$. Therefore the $\alpha_{j,j+1}$ root string through $\alpha_{i,i+1}$ contains a second vector iff $j = i + 1$; and the $\alpha_{i,i+1}$ root string through $\alpha_{j,j+1}$ contains a second vector iff $i = j + 1$. Hence $\alpha_{i,i+1}$ and $\alpha_{j,j+1}$ are joined iff $|i - j| = 1$, and in that case only by one edge. The Dynkin diagram is thus

$$\underset{\alpha_{12}}{\bullet} \rule[0.4ex]{2em}{0.4pt} \underset{\alpha_{23}}{\bullet} \cdots \underset{\alpha_{l-1,l}}{\bullet} \rule[0.4ex]{2em}{0.4pt} \underset{\alpha_{l,l+1}}{\bullet}.$$

Before beginning our discussion of the three other classical families we introduce some notation. Let $a_+ = \left(\begin{smallmatrix} 0 & 1 \\ 0 & 0 \end{smallmatrix}\right)$, $a_- = \left(\begin{smallmatrix} 0 & 0 \\ 1 & 0 \end{smallmatrix}\right)$ and $a_0 = \left(\begin{smallmatrix} 1 & 0 \\ 0 & -1 \end{smallmatrix}\right)$. We consider $n \times n$ matrices whose entries are 2×2 matrices. Thus the resulting matrix is actually $2n \times 2n$. For example, consider the 4×4 matrix given by

$$\begin{pmatrix} 0 & 0 & 0 & 1 \\ 0 & 0 & 0 & 0 \\ 0 & 0 & 0 & 0 \\ 0 & 0 & 0 & 0 \end{pmatrix} = \begin{pmatrix} 0 & a_+ \\ 0 & 0 \end{pmatrix}.$$

In the matrix on the right the zeroes denote the 2×2 zero matrix. We shall denote the above matrix by $(a_+)_{12}$ to indicate that the matrix a_+ is placed in the $(1,2)$ slot of a 2×2 matrix. Such matrices are often called *super matrices*. The classical algebras B_l, C_l, and D_l are conveniently expressed in terms of such matrices. The series B_l is slightly more complicated than the other two, so we shall discuss it last.

C_l. This is the algebra $sp(2l, C)$ of $2l \times 2l$ symplectic matrices B such that $BJ + JB' = 0$ where $J = \mathrm{diag}(a, a, \ldots, a)$ with $a = a_+ - a_-$. We represent each matrix in $sp(2l, C)$ by an $l \times l$ supermatrix each of whose entries are 2×2 matrices. The Cartan subalgebra consists of matrices $H = \mathrm{diag}(\lambda_1 a_0, \lambda_2 a_0, \ldots, \lambda_l a_0)$. We define the functionals α_i on $\mathfrak{h}'$ by $\alpha_i(H) = \lambda_i$. The roots and root vectors are as follows:

root vectors	roots	
$(a_+)_{ij} + (a_+)_{ji}$	$\alpha_i + \alpha_j$	$i < j$
$(a_-)_{ij} + (a_-)_{ji}$	$-(\alpha_i + \alpha_j)$	$j < i$
$(a_+)_{ii}$	$2\alpha_i$	
$(a_-)_{ii}$	$-2\alpha_i$	
$\left(\dfrac{\mathbb{1} + a_0}{2}\right)_{ij} - \left(\dfrac{\mathbb{1} - a_0}{2}\right)_{ji}$	$\alpha_i - \alpha_j$	$i < j$
$\left(\dfrac{\mathbb{1} - a_0}{2}\right)_{ij} - \left(\dfrac{\mathbb{1} + a_0}{2}\right)_{ji}$	$-(\alpha_i - \alpha_j)$	$i < j$

(10.6)

Define the positive roots to be those which are positive on $H_0 = \mathrm{diag}(\lambda_1 a_0, \ldots, \lambda_l a_0)$ with $\lambda_1 > \cdots > \lambda_l > 0$. Then the positive roots are $\alpha_i - \alpha_j$ for $i < j$,

$\alpha_i + \alpha_j$, and $2\alpha_i$. The fundamental roots are $\alpha_1 - \alpha_2, \alpha_2 - \alpha_3, \ldots, \alpha_{l-1} - \alpha_l$, $2\alpha_l$. Again, the simple roots $\alpha_{i,i+1}$ and $\alpha_{j,j+1}$ are linked iff $|i - j| = 1$; otherwise not. At the end, we have

$$2\alpha_l + (\alpha_{l-1} - \alpha_l) = \alpha_{l-1} + \alpha_l$$

$$(\alpha_{l-1} + \alpha_l) + (\alpha_{l-1} - \alpha_l) = 2\alpha_{l-1};$$

so the $\alpha_{l-1} - \alpha_l$ root string through $2\alpha_l$ contains three elements. This means that $2\alpha_l$ is the longer root and that the vertices for $2\alpha_l$ and $\alpha_{l-1} - \alpha_l$ are joined by two edges. The Dynkin diagram is

$$\underset{\alpha_1 \ - \ \alpha_2}{\bullet \underline{\hspace{1.5cm}} \bullet} \ \underset{\alpha_2 \ - \ \alpha_3}{} \quad \cdots \ \underset{\alpha_{l-1} \ - \ \alpha_l}{\bullet \Longrightarrow} \underset{2\alpha_l.}{\bullet}$$

D_l. This is the series $so(2l, C)$. It is the Lie algebra of the matrix group which preserves the quadratic form $x_1^2 + \cdots + x_{2l}^2$. By making a unitary transformation on $\mathbb{R}^{2l}$ this quadratic form is changed into the form $2(w_1 w_2 + w_3 w_4 + \cdots + w_{2l-1} w_{2l})$. (Specifically, put $w_1 = (x_1 + ix_2)/\sqrt{2}$, $w_2 = (x_1 - ix_2)/\sqrt{2}$, etc.) Therefore $so(2l, C)$ is isomorphic (under a complex but not a real isomorphism) to the algebra of matrices B such that $BG + GB^t = 0$ where $G = \text{diag}(a, a, \ldots, a)$, with $a = a_+ + a_- = \begin{pmatrix} 0 & 1 \\ 1 & 0 \end{pmatrix}$. Again we take the Cartan subalgebra to be matrices $H = \text{diag}(\lambda_1 a_0, \lambda_2 a_0, \ldots, \lambda_l a_0)$. The roots and root vectors are

$$\begin{array}{lll} (a_-)_{ij} - (a_-)_{ji} & -(\alpha_i + \alpha_j) & i < j \\[2mm] (a_+)_{ij} - (a_+)_{ji} & \alpha_i + \alpha_j & j < i \\[2mm] \left(\dfrac{1 + a_0}{2}\right)_{ij} + \left(\dfrac{a_0 - 1}{2}\right)_{ji} & \alpha_i - \alpha_j & i \neq j. \end{array} \qquad (10.7)$$

The ordering is again determined by $H_0 = (\lambda_1 a_0, \ldots, \lambda_l a_0)$ with $\lambda_1 > \cdots > \lambda_l > 0$. Then the positive roots are $\alpha_i - \alpha_j$ for $i < j$ and $\alpha_i + \alpha_j$. The fundamental roots are $\alpha_1 - \alpha_2, \alpha_2 - \alpha_3, \ldots, \alpha_{l-1} - \alpha_l$, and $\alpha_{l-1} + \alpha_l$; and the Dynkin diagram is

$$\underset{\alpha_1 \ - \ \alpha_2}{\bullet \underline{\hspace{1.5cm}} \bullet} \ \underset{\alpha_2 \ - \ \alpha_3}{} \quad \cdots \ \bullet \diagdown\!\!\!\diagup \begin{array}{l} \bullet \ \alpha_{l-1} - \alpha_l \\[2mm] \bullet \ \alpha_{l-2} - \alpha_{l-1}. \\[2mm] \bullet \ \alpha_{l-1} + \alpha_l \end{array}$$

B_l. We finally come to the algebras $so(2l + 1, C)$, the odd dimensional orthogonal algebras. We again make a unitary transformation in $\mathbb{R}^{2l+1}$ as follows. Let the coordinates in $\mathbb{R}^{2l+1}$ be $(x_0, x_1 \ldots x_{2l})$ and put $w_0 = x_0$, $w_1 = (x_1 + ix_2)/\sqrt{2}$, $w_2 = (x_1 - ix_2)/\sqrt{2}$, etc. Then $x_0^2 + x_1^2 + \cdots + x_2^2 = w_0^2 + 2(w_1 w_2 + w_3 w_4 + \cdots)$; and $so(2l + 1, C)$ is (complex) isomorphic to the algebra of matrices B such that $BG + GB^t = 0$, where

$$G = \begin{pmatrix} \boxed{1} & & & & \\ & \boxed{a} & & & \\ & & \boxed{a} & & \\ & & & \ddots & \\ & & & & \boxed{a} \end{pmatrix}.$$

The Cartan subalgebra consists of matrices $H = (0, \lambda_1 a_0, \lambda_2 a_0, \ldots, \lambda_l a_0)$. Now our supermatrices are $(l + 1) \times (l + 1)$ matrices with scalars in the first row and column (labelled by $(0, i)$ and $(i, 0)$) and 2×2 matrices in the ijth slot for $1 \le i, j \le l$. Define the functionals α_i by $\alpha_i(H) = \lambda_i$. The roots of B_l are

root vectors	*roots*	
$E_{0,2i} - E_{2i-1,0}$	α_i	$1 \le i \le l$
$E_{2i,0} - E_{0,2i-1}$	$-\alpha_i$	
$\left(\dfrac{1+a_0}{2}\right)_{ij} - \left(\dfrac{1-a_0}{2}\right)_{ji}$	$\alpha_i - \alpha_j$	$\begin{array}{l} 1 \le i, j \le l \\ i \neq j \end{array}$
$\left.\begin{array}{l}(a_-)_{ij} - (a_-)_{ji} \\ (a_+)_{ij} - (a_+)_{ji}\end{array}\right\}$	$\alpha_i + \alpha_j$	$i < j$
	$-(\alpha_i + \alpha_j)$	

$$(10.8)$$

An order is defined by H_0 with $\lambda_1 > \lambda_2 > \cdots > \lambda_l > 0$. The positive roots are then α_i, $\alpha_i + \alpha_j$, and $\alpha_i - \alpha_j$ with $i < j$. The fundamental roots are $\alpha_1 - \alpha_2, \ldots$, $\alpha_{l-1} - \alpha_l$, and α_l, and the Dynkin diagram is

$$\underset{\alpha_1 - \alpha_2}{\bullet\!\!-\!\!\bullet} \ \underset{\alpha_2 - \alpha_3}{} \ \cdots \ \underset{\alpha_{l-1} - \alpha_l}{\bullet} \!\Longleftarrow\! \underset{\alpha_l}{\bullet}.$$

$$B_l$$

We now turn to the computation of the Weyl groups of the classical algebras. Let α and β be roots and let S_α be the Weyl reflection corresponding to α. Then $S_\alpha(\alpha) = -\alpha$ and $S_\alpha(\beta) = \beta - a_{\beta\alpha}\alpha$ where $a_{\beta\alpha}$ is the Cartan integer. By Lemma 10.7(i) $a_{\beta\alpha} = -(p + q)$ where $p \le 0$ and $q \ge 0$ are the limits of the α root string through β. Thus $S_\alpha(\beta) = \beta + (p + q)\alpha$ where $\beta + n\alpha$ are roots for $p \le n \le q$.

Let us first find the Weyl group for the algebra A_l. The roots are $\alpha_{ij} = \alpha_i - \alpha_j$. Let S_{12} be the Weyl reflection corresponding to the root α_{12}. Then $S_{12}(\alpha_{12}) = -\alpha_{12} = \alpha_2 - \alpha_1$. To find $S_{12}(\alpha_{2m})$ note that the α_{12} root string through α_{2m} is α_{2m}, $\alpha_{2m} + \alpha_{12} = \alpha_{1m}$; hence $p = 0$, $q = 1$, and $S_{12}(\alpha_{2m}) = \alpha_{2m} + \alpha_{12} = \alpha_{1m}$. Similarly $S_{12}(\alpha_{mn}) = \alpha_{mn}$ if neither m nor n is equal to 1 or 2. Thus S_{12} acts by interchanging α_1 and α_2—that is, as a transposition of the 1-2 coordinates. The collection of all such transpositions S_{ij} generates the permutation group S_{l+1}; this is the Weyl group for A_l.

Now let us look at B_l. The roots are $\pm\alpha_i$, $\alpha_i - \alpha_j$ for $i \neq j$, and $\pm(\alpha_i + \alpha_j)$ for $i \neq j$. The reflection S_i corresponding to α_i takes α_i to $-\alpha_i$. What does it do to α_j? The α_i string through α_j is $\alpha_j - \alpha_i$, α_j, $\alpha_j + \alpha_i$; so $p = -1$, $q = 1$, and

$S_i(\alpha_j) = \alpha_j$. The Weyl group thus contains all the reflections $\alpha_i \to -\alpha_i$. Now let S_{ij} be the reflection corresponding to the root $\alpha_i - \alpha_j$. We find by our root string algorithm that $S_{ij}(\alpha_i) = \alpha_j$, $S_{ij}(\alpha_j) = \alpha_i$, and $S_{ij}(\alpha_m) = \alpha_m$ if $m \neq i$ or j. For example, the α_{ij} root string through α_i is $\alpha_i - (\alpha_{ij})$, α_i; hence $q = 0$, $p = -1$, and $S_{ij}(\alpha_j) = \alpha_j + (\alpha_i - \alpha_j) = \alpha_i$. Finally, no new operations are obtained from the Weyl reflections corresponding to the $\pm(\alpha_i + \alpha_j)$. For example the Weyl reflection corresponding to the root $\alpha_i + \alpha_j$ interchanges α_i and $-\alpha_j$. This is a combination of a transposition and a reflection. Thus the Weyl group for B_l contains all permutations and all reflections α_i to $-\alpha_i$. Its order is $2^l l!$

The roots of C_l are the same as those for B_l except that $\pm\alpha_i$ are replaced by $\pm 2\alpha_i$. One finds that C_l has the same Weyl group as B_l.

The roots of D_l are $\alpha_i - \alpha_j$ and $\pm(\alpha_i + \alpha_j)$. As before, the reflections S_{ij}^- corresponding to the roots $(\alpha_i - \alpha_j)$ are the transpositions $\alpha_i \leftrightarrow \alpha_j$. Let us consider the reflection S_{ij}^+ corresponding to the root $\alpha_i + \alpha_j$. We find easily that

$$S_{ij}^+(\alpha_i - \alpha_j) = (\alpha_i - \alpha_j)$$
$$S_{ij}^+(\alpha_i + \alpha_j) = -(\alpha_i + \alpha_j).$$

Therefore $S_{ij}^+(\alpha_i) = -\alpha_j$ and $S_{ij}^+(\alpha_j) = -\alpha_i$. Similarly $S_{ij}^+(\alpha_m) = \alpha_m$ if $m \neq i, j$. The combined operation $S_{ij}^- S_{ij}^+$ acts by changing α_i to $-\alpha_i$, α_j to $-\alpha_j$ and leaving all other coordinates fixed. Thus the Weyl group for D_l contains all permutations and all even sign changes, that is, reflections of an even number of coordinates. Its order is $2^{l-1} l!$

EXERCISES

1. Let $\omega_1, \ldots \omega_l$ be the simple roots for one of the classical algebras A_l, B_l, C_l, D_l. Compute the corresponding elements $T_i \in \mathfrak{h}$ given by $[E_i, E_{-i}] = T_i$, where E_i and E_{-i} are the root vectors in the presentations of these algebras given in this section. Show that $[T_i, E_j] = a_{ij} E_j$, where a_{ij} is the Cartan matrix; and that $[E_i, E_j] = 0$ for $i + j \neq 0$.

2. Using the results in exercise 1, find functionals λ_i in $\mathfrak{h}'$ such that $\lambda_i(T_j) = \delta_{ij}$. Show they are given as follows:

A_l: $\lambda_1 = \alpha_1$, $\quad \lambda_2 = \alpha_1 + \alpha_2$, $\ldots$ $\lambda_l = \alpha_1 + \alpha_2 + \cdots + \alpha_l$

B_l: $\lambda_1 = \alpha_1$, $\quad \lambda_2 = \alpha_1 + \alpha_2$, $\ldots$ $\lambda_{l-1} = \alpha_1 + \alpha_2 + \cdots + \alpha_{l-1}$

$\quad \lambda_l = (\tfrac{1}{2})(\alpha_1 + \cdots + \alpha_1)$

C_l: $\lambda_1 = \alpha_1$, $\quad \lambda_2 = \alpha_1 + \alpha_2$, $\ldots$ $\lambda_l = \alpha_1 + \alpha_2 + \cdots + \alpha_l$

D_l: $\lambda_1 = \alpha_1$, $\quad \lambda_2 = \alpha_1 + \alpha_2$, $\ldots$ $\lambda_{l-2} = \alpha_1 + \alpha_2 + \cdots + \alpha_{l-2}$

$\quad \lambda_{l-1} = (\tfrac{1}{2})(\alpha_1 + \cdots + \alpha_{l-1} - \alpha_l)$

$\quad \lambda_l = (\tfrac{1}{2})(\alpha_1 + \cdots + \alpha_{l-1} + \alpha_l).$

These functionals are *weights* for the fundamental representations of the classical algebras. This topic will be discussed in detail in §40.

3. Obtain a realization of the Lie algebra $sp(2l, C)$ by boson creation and annihilation operators. (See 4.7.) *Hint*: $[a_i a_i^*, a_i a_j] = -a_i a_j$. Elements of the Cartan subalgebra are given by $H = \sum_i \lambda_i a_i a_i^*$.

4. Obtain a realization of the Lie algebra $so(2l)$ by Fermion creation and annihilation operators. These satisfy the anticommutation relations

$$\{a_i, a_j\} = \{a_i^*, a_j^*\} = 0, \qquad \{a_i, a_j^*\} = \delta_{ij} \mathbb{1},$$

where $\{A, B\} = AB + BA$.

CHAPTER 11

Real Forms

37. Compact Real Forms; Weyl's Theorem

In the last chapter we classified all the semi-simple algebras over the complex numbers; but each complex algebra in general has several real forms. For example, we saw in Chapter 1 that $sl(2, R)$ and $su(2)$ are distinct real forms of the complex algebra $sl(2, C)$. That is, there is no real isomorphism between them, yet they have the same complexification. The task of classifying the real forms of the semi-simple algebras is more involved than that of classifying the complex forms themselves. It was first done by Cartan in 1914. Later, in 1926–28 he used this classification to determine all the symmetric spaces. A complete discussion of the real forms and their relationship to the symmetric spaces is given in the well known book of Helgason, and we shall give only a brief introduction here.

Let $\mathfrak{g}$ be a complex Lie algebra. A *real basis* for $\mathfrak{g}$ is one for which all the structure constants are real. In the last chapter we saw that every complex simple algebra in fact has a real basis, namely the Weyl–Chevalley basis. If $X_1, \ldots X_n$ is a real basis for $\mathfrak{g}$ we can form a real Lie algebra $\mathfrak{r}$ by taking all real linear combinations of the X_i. The resulting real algebra $\mathfrak{r}$ is called a *real form* of $\mathfrak{g}$: its complexification is $\mathfrak{g}$. More generally, we say that $\mathfrak{r}$ is a real form of the complex algebra $\mathfrak{g}$ if $\mathfrak{r}$ is a real Lie algebra whose complexification is $\mathfrak{g}$. Note that the matrices of $\mathfrak{r}$ need not themselves be real.

For example, the matrices L_1, L_2, L_3 given by (2.1) generate the real algebra $so(3, R)$. The complexification of this algebra is clearly $so(3, C)$, the algebra of 3×3 complex skew symmetric matrices. Now $iL_1, iL_2,$ and L_3 also have real commutation relations, *viz.*

$$[iL_1, iL_2] = -L_3 \qquad [iL_2, L_3] = iL_1 \qquad [L_3, iL_1] = iL_2$$

and therefore they also generate a real Lie algebra. The complexification of that algebra is also $so(3, C)$. On the other hand, it is easily checked that $\{iL_1, iL_2, L_3\}$ is isomorphic to the real algebra, $so(2, 1)$ of all real matrices R such that $Rg + gR^t = 0$, where $g = \mathrm{diag}(1, -1, -1)$.

Now let us prove that the real algebras $so(3, R)$ and $so(2, 1)$ are distinct real forms, that is, that there is no real isomorphism between them. The Killing forms of isomorphic algebras are necessarily identical, since an isomorphism preserves the structure constants. Let K be the Killing form of a real algebra $\mathfrak{g}$ relative to the basis $\{X_i\}$. Let $Z_i = \sum_j O_{ij} X_j$. Then $K(Z_m, Z_n) = K(O_{mr} X_r, O_{ns} X_s) = O_{mr} O_{ns} K(X_r, X_s)$; in other words, the metric tensor g is transformed into OgO^t. Since K is symmetric, we can diagonalize it by a real orthogonal matrix O. Therefore a necessary condition that two real forms be isomorphic is that their Killing forms have the same signature. In the case of the algebras $so(3, R)$ and $so(2, 1)$, however, the Killing forms are

$$K = \begin{pmatrix} -2 & 0 & 0 \\ 0 & -2 & 0 \\ 0 & 0 & -2 \end{pmatrix} \qquad K = \begin{pmatrix} 2 & 0 & 0 \\ 0 & 2 & 0 \\ 0 & 0 & -2 \end{pmatrix}$$

respectively, and so there can be no real isomorphism between them.

A *compact real form* of an algebra $\mathfrak{g}$ is one for which the Killing form is negative definite. The reason for this definition is the following important theorem, due to Weyl:

Theorem 11.1 (Weyl). *A (real) Lie group is compact if and only if its Killing form is negative definite.*

PROOF. Recall (Theorem 7.1) that a compact Lie group has a bi-invariant measure, say $d\mu$. Let ρ be any representation of $\mathfrak{G}$ on a vector space V. By integrating over the group we can construct a positive definite Hermitian inner product on V which is invariant under the representation ρ. In fact, if $((\ ,\ ))$ is any scalar product on V we set

$$(u, v) = \int_{\mathfrak{G}} ((\rho(g)u, \rho(g)v))\, d\mu(g).$$

It is easily seen that $(\rho(h)u, \rho(h)v) = (u, v)$; hence ρ is a representation of $\mathfrak{G}$ by unitary matrices. This means that the infinitesimal generators are skew Hermitian: that is, $\rho(X)^* = -\rho(X)$ for any X in $\mathfrak{g}$. (Here we are using ρ to denote both the representation of the group and the algebra.) It follows that $\rho(X)$ is diagonalizeable for each X in $\mathfrak{g}$ and that its eigenvalues are purely imaginary. Hence $\mathrm{Tr}\,\rho(X)^2 < 0$. Applying this result to the adjoint representation we find that the Killing form is negative definite. More generally, the above arguments apply to the representations of any compact Lie group.

Conversely, suppose the Killing form is negative definite. Let $\mathrm{Aut}\,\mathfrak{g}$ be the group of automorphisms of $\mathfrak{g}$. Each $\sigma \in \text{Aut } \mathfrak{g}$ preserves the Killing form. If we represent σ by a matrix R, then we have $RKR^t = K$. Since K is symmetric and negative definite, the set of such matrices R is isomorphic to a subgroup of orthogonal matrices, hence is compact. Now there is a natural homomorphism of $\mathfrak{G}$ into Aut $\mathfrak{g}$ given by $g \to \sigma_g$, where $\sigma_g(X) = gXg^{-1}$. The image under this homomorphism is denoted by Ad $\mathfrak{G}$. Thus $\mathfrak{G}/\ker \sigma \cong \text{Ad } \mathfrak{G}$ and is compact. The kernel of σ is the center of $\mathfrak{G}$ and is a closed subgroup. In order to show that $\mathfrak{G}$ is compact, one must show that $\ker \sigma$ is actually finite. This is the deep part of Weyl's theorem. We omit this step. (See Varadarajan, p. 345; a different proof appears in Helgason, p. 123.) $\square$

Theorem 11.2. *Every semi-simple Lie algebra has a compact real form.*

PROOF. Beginning with the Weyl–Chevalley basis we choose as a basis

$$\{iH_\alpha, \quad X_\alpha = E_\alpha - E_{-\alpha}, \quad Y_\alpha = i(E_\alpha + E_{-\alpha})\}.$$

The operators E_α were normalized so that $[E_\alpha, E_{-\alpha}] = H_\alpha$, so from Lemma 10.5, (iii), $K(E_\alpha, E_{-\alpha}) = 1$, $K(E_\alpha, E_\alpha) = 0$; and

$$K(X_\alpha, X_\alpha) = -2, \qquad K(Y_\alpha, Y_\alpha) = -2, \qquad K(X_\alpha, Y_\beta) = 0.$$

Moreover, $[iH_\alpha, X_\beta] = \beta(H_\alpha)Y_\beta$ and $[iH_\alpha, Y_\beta] = -\beta(H_\alpha)X_\beta$; so

$$K(iH_\alpha, iH_\alpha) = \sum_\beta (\beta(H_\alpha)^2 < 0.$$

It follows that K is negative definite for this basis. We must still check that this is a real basis for $\mathfrak{g}$. For this one must use the normalization $N_{\alpha\beta} = -N_{(-\alpha)(-\beta)}$.

Finally, we close this section with an important application of Weyl's theorem concerning the reducibility of representations of semi-simple Lie algebras. (For a purely algebraic proof, see Samelson p. 108 or Jacobson.) $\square$

Theorem 11.3. *Every finite dimensional representation ρ of a semi-simple algebra $\mathfrak{g}$ decomposes into the direct sum of irreducible representations.*

PROOF. Let $\mathfrak{u}$ be a compact real form of $\mathfrak{g}$; and let $\mathfrak{U}$ be the corresponding compact real Lie group. As in the proof of Theorem 11.1 we construct an invariant positive definite Hermitian symmetric form on the representation space V by integrating with respect to the invariant measure over $\mathfrak{U}$. The matrices $\rho(X)$, $X \in \mathfrak{u}$ are then skew Hermitian with respect to this inner product. If ρ is reducible, then there exists a subspace V_1 of V which is invariant under $\rho(X)$ for all $X \in \mathfrak{u}$. Since $\rho(X)^* = -\rho(X)$, the orthogonal complement of V_1 is also invariant under the representation. If these spaces are in turn reducible, we may repeat the procedure until the entire space is decomposed. $\square$

38. Cartan Decompositions

In Chapter 1 we saw that $SL(2, R)$ was the isometry group of the upper half plane; moreover $SO(2)$ is a subgroup of $SL(2, R)$ that leaves the point i fixed. (Recall that $SL(2, R)$ acts on the upper half plane as a group of Möbius transformations.) The quotient space $SL(2, R)/SO(2)$ is in one to one correspondance with the upper half plane. (See also exercise 8, § 4 for the corresponding construction for the unit disk.) This situation illustrates a general and important construction of *symmetric spaces*.

Let us look at what happens on the Lie algebra level. The algebra $sl(2, R)$ is a direct sum of the vector spaces:

$$\mathfrak{k} = sp\left\{\begin{pmatrix} 0 & 1 \\ -1 & 0 \end{pmatrix}\right\}, \quad \text{and} \quad \mathfrak{p} = sp\left\{\begin{pmatrix} 0 & 1 \\ 1 & 0 \end{pmatrix}, \begin{pmatrix} 1 & 0 \\ 0 & -1 \end{pmatrix}\right\}$$

and we may write $sl(2, R) = \mathfrak{k} + \mathfrak{p}$ where

$$[\mathfrak{k}, \mathfrak{k}] \subseteq \mathfrak{k}, \qquad [\mathfrak{k}, \mathfrak{p}] \subseteq \mathfrak{p}, \qquad [\mathfrak{p}, \mathfrak{p}] \subseteq \mathfrak{k}. \tag{11.1}$$

This is the *Cartan decomposition* of $sl(2, R)$. We have

Theorem 11.4 (Cartan Decomposition). *Every semi-simple real Lie algebra $\mathfrak{g}$ has a direct sum decomposition of the form $\mathfrak{g} = \mathfrak{k} + \mathfrak{p}$ where $\mathfrak{k}$ and $\mathfrak{p}$ satisfy (11.1). The Killing form is negative definite on $\mathfrak{k}$ and positive definite on $\mathfrak{p}$; and $\mathfrak{k}$ and $\mathfrak{p}$ are orthogonal in the metric tensor. The decomposition is unique up to an inner automorphism of $\mathfrak{g}$ by an element of the group.*

The existence of a Cartan decomposition for any real form will be given in the course of the proof of Theorem 11.5 below.

The Cartan decomposition of $sl(2, R)$ could have been found in the following way. Consider the compact Lie algebra $su(2)$, generated by the matrices

$$E_1 = (\tfrac{1}{2})\begin{pmatrix} 0 & -i \\ -i & 0 \end{pmatrix} \qquad E_2 = (\tfrac{1}{2})\begin{pmatrix} 0 & -1 \\ 1 & 0 \end{pmatrix} \qquad E_3 = (\tfrac{1}{2})\begin{pmatrix} -i & 0 \\ 0 & i \end{pmatrix}.$$

Let τ be the operation of complex conjugation. It is easily seen that τ is an automorphism of the real algebra $su(2)$. Since $\tau^2 = 1$ its eigenvalues are ± 1. The subalgebra $\mathfrak{k}$ is obtained as the $+1$ eigenspace of τ and the subspace $\mathfrak{p}$ is obtained as the -1 eigenspace of τ. Again, this is the general case. We have (Cartan)

Theorem 11.5. *Let $\mathfrak{u}$ be a compact real form of the Lie algebra $\mathfrak{g}$ and let σ be an involutory ($\sigma^2 = 1$) automorphism of $\mathfrak{u}$. Then $\mathfrak{r} = \mathfrak{k} + i\mathfrak{p}$ is a real form, where $\mathfrak{k}$ and $\mathfrak{p}$ are resepectively the $+1$ and -1 eigenspaces of σ; and $\mathfrak{k}$ and $i\mathfrak{p}$ are its Cartan decomposition.*

Conversely, given any real form $\mathfrak{r}$ there exists a compact form $\mathfrak{w}$ and an automorphism σ of $\mathfrak{w}$ such that $\mathfrak{r} = \mathfrak{k} + i\mathfrak{p}$, where $\mathfrak{k}$ and $\mathfrak{p}$ are the $+1$ and -1 eigenspaces of σ.

PROOF. Since $\sigma^2 = 1$, its only eigenvalues are 1 and -1. The inclusion relations (11.1) follow immediately from $\sigma[X, Y] = [\sigma X, \sigma Y]$. Since σ is an automorphism of $\mathfrak{u}$ it preserves the Killing form: $K(\sigma X, \sigma Y) = K(X, Y)$. Moreover, since $\mathfrak{u}$ is a compact real form, K is negative definite everywhere. Hence K is negative definite on $\mathfrak{k}$ and positive definite on $i\mathfrak{p}$. To show that $\mathfrak{k}$ and $\mathfrak{p}$ are orthogonal with respect to K, note that for $X \in \mathfrak{k}$ and $Y \in \mathfrak{p}$ we have $K(X, Y) = K(\sigma X, \sigma Y) = K(X, -Y) = -K(X, Y)$; and therefore $K(X, Y) = 0$. Finally, it is easily seen that the inclusions (11.1) also hold for $\mathfrak{k}$ and $i\mathfrak{p}$; so $\mathfrak{k} + i\mathfrak{p}$ is a real subalgebra.

Conversely, let $\mathfrak{r}$ be a real form of $\mathfrak{g}$. Then every element of $\mathfrak{g}$ can be uniquely written in the form $U + iV$, where U and V belong to $\mathfrak{r}$. We define σ on $\mathfrak{g}$ by $\sigma(U + iV) = U - iV$ and call σ the *conjugation of $\mathfrak{g}$ with respect to $\mathfrak{r}$*. It is obvious that $\sigma^2 = 1$ and easily seen that σ is an automorphism of $\mathfrak{g}$. Similarly, there is a conjugation of $\mathfrak{g}$ with respect to any compact real form $\mathfrak{u}$, this automorphism being denoted by τ. The Killing form of $\mathfrak{g}$ may be obtained either as an extension of the Killing form on $\mathfrak{r}$ or $\mathfrak{u}$ to vectors with complex coefficients. Since K is real on $\mathfrak{r}$ or $\mathfrak{u}$ we find that $K(\sigma X, \sigma Y) = K(X, Y)^* = K(\tau X, \tau Y)$ for any X and Y in $\mathfrak{g}$.

A Hermitian symmetric inner product on $\mathfrak{g}$ is defined by $\langle X, Y \rangle = K(X, \tau Y)$: we have $\langle Y, X \rangle = K(Y, \tau X) = K(\tau X, Y) = K(X, \tau Y)^* = \langle X, Y \rangle^*$. Moreover, $\langle X, X \rangle < 0$ for $X \in \mathfrak{g}$: simply put $X = U + iV$, where U and V are in $\mathfrak{u}$; then $\langle U + iV, U + iV \rangle = K(U + iV, U - iV) = K(U, U) + K(V, V) < 0$, since K is negative definite on the compact form $\mathfrak{u}$.

Since σ and τ are both automorphisms of $\mathfrak{g}$, so is $\sigma\tau$. Denote this automorphism by ω. We claim that ω is self adjoint with respect to $\langle \, , \, \rangle$. In fact, $\langle \omega X, Y \rangle = K(\sigma\tau X, \tau Y) = K(\tau X, \sigma\tau Y)^* = K(X, \tau\sigma\tau Y) = \langle X, \omega Y \rangle$. It follows that the eigenvalues of ω are real and that $\mathfrak{g}$ decomposes into a direct sum of eigenspaces of ω. Denote the eigenvalues by λ_i and the corresponding eigenspaces by $\mathfrak{g}_i$. Writing $\omega = \operatorname{diag}(\ldots \lambda_i \ldots)$, we define f_ω to be the operator $\operatorname{diag}(\ldots f(\lambda_i) \ldots)$ for any real valued function f. The operators ω and f_ω are simultaneously diagonalized and so commute.

We claim that for any f satisfying $f(xy) = f(x)f(y)$, f_ω is an automorphism of $\mathfrak{g}$, provided that f does not vanish on any of the eigenvalues of ω. Since ω preserves the brackets, it follows that $[\mathfrak{g}_i, \mathfrak{g}_j]$ is either 0 or lies in the eigenspace with eigenvalue $\lambda_i\lambda_j$. Therefore for $X_i \in \mathfrak{g}_i$ and $X_j \in \mathfrak{g}_j$ we have $f_\omega[X_i, X_j] = f(\lambda_i\lambda_j)[X_i, X_j] = f(\lambda_i)f(\lambda_j)[X_i, X_j] = [f(\lambda_i)X_i, f(\lambda_j)X_j] = [f_\omega X_i, f_\omega X_j]$.

In particular this is true for $f(\lambda) = |\lambda|^{1/2}$ and $\operatorname{sgn}\lambda$, since none of the eigenvalues λ_i vanish. Denote these automorphisms by μ and ν respectively; then $\omega = \mu^2\nu$.

Now $\tau\omega\tau^{-1} = \omega^{-1}$ and so $\tau f_\omega \tau^{-1} = f_{\omega^{-1}} = (f_\omega)^{-1}$ for any f. In particular, $\tau\mu\tau^{-1} = \mu^{-1}$ and $\tau\nu\tau^{-1} = \nu$, hence $\tau\nu = \nu\tau$. Moreover, μ and ν commute. From these relationships we find that $\mu^{-1}\sigma\mu = \mu^{-1}\omega\tau\mu = \mu^{-1}\mu^2\nu\tau\mu = \mu\nu\tau\mu = \mu\nu\mu^{-1}\tau = \nu\tau$.

We define $J = \nu\tau$. Since ν and τ commute, J commutes with τ and, considered as an automorphism of $\mathfrak{g}$, it leaves $\mathfrak{u}$ invariant. Let $\mathfrak{w} = \mu(\mathfrak{u})$; since μ

is an automorphism, $\mathfrak{w}$ is also a compact real form; and since $J = \mu^{-1}\sigma\mu$ we find that σ is an involutory automorphism of $\mathfrak{w}$:

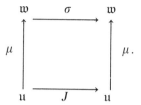

Let $\mathfrak{k}$ and $\mathfrak{p}$ be the $+1$ and -1 eigenspaces of σ, so that $\mathfrak{w} = \mathfrak{k} + \mathfrak{p}$. We are going to prove that $\mathfrak{r} = \mathfrak{k} + i\mathfrak{p}$. Since $\mathfrak{w}$ is a compact form, the Killing form K must be negative definite on $\mathfrak{k}$ and $\mathfrak{p}$. By the argument in the first part of the theorem, $\mathfrak{k} + i\mathfrak{p}$ is also a real form of $\mathfrak{g}$, and $\mathfrak{k}$ and $i\mathfrak{p}$ are its Cartan decomposition.

Recall that σ was the conjugation with respect to $\mathfrak{r}$. So if X and Y are in $\mathfrak{r}$ then $\sigma(X + iY) = X - iY$. Since $\sigma X = X$ iff $X \in \mathfrak{k}$, we find that $\mathfrak{k} \subseteq \mathfrak{r}$. Now we claim that $i\mathfrak{p} \subseteq \mathfrak{r}$. Say $X \in \mathfrak{p}$ and write $X = U + iV$, where U and $V \in \mathfrak{r}$. Then $\sigma(X) = -X = -(U + iV)$ since $X \in \mathfrak{p}$; and $\sigma(X) = U - iV$ since σ was conjugation with respect to $\mathfrak{r}$. It follows that $U = 0$ and $X = iV$, where $V \in \mathfrak{r}$, hence $iX = -V \in \mathfrak{r}$. Thus both $\mathfrak{k}$ and $i\mathfrak{p}$ are contained in $\mathfrak{r}$. Since these two subspaces are orthogonal they are linearly independent and so by dimensional arguments they span $\mathfrak{r}$. This completes the proof of Theorem 11.5. □

For example, $\sigma(A) = -A^t$ is an involutory automorphism of $sl(n, R)$. It is immediate that $\mathfrak{k}$ consists of the skew symmetric $n \times n$ matrices and $\mathfrak{p}$ consists of the symmetric matrices. Thus the Cartan decomposition of $sl(n, R)$ is the familiar decomposition of a matrix into its symmetric and antisymmetric parts. Moreover, the dual space $\mathfrak{k} + i\mathfrak{p}$ space is precisely the algebra $su(n)$; for it consists of matrices of the form $U + iV$ where $U^t = -U$ and $V^t = V$. The corresponding symmetric spaces are $SL(n, R)/SO(n, R)$ and $SU(n)/SO(n, R)$. The first of these is a noncompact Riemannian space of positive curvature, and the second is a compact Riemannian space of negative curvature. See Helgason.

Recall that the algebra $so(p, q)$ consists of all matrices U such that $U^t I_{p,q} + I_{p,q}U = 0$, where

$$I_{p,q} = \begin{pmatrix} -I_p & 0 \\ 0 & I_q \end{pmatrix}$$

where I_p is the $p \times p$ identity matrix. It is easily seen that the matrices in $so(p, q)$ take the form

$$U = \begin{pmatrix} A & B \\ B^t & C \end{pmatrix}$$

where A and C are skew symmetric matrices of size p and q respectively, and

B is any real $p \times q$ matrix. We define the automorphism $\sigma_{p,q}$ of $so(p,q)$ to be $\sigma_{p,q}(U) = I_{p,q} U I_{p,q}$. Then $\sigma_{p,q}^2 = 1$ and

$$\mathfrak{k} = \left\{ \begin{pmatrix} A & 0 \\ 0 & C \end{pmatrix} \middle| A, C \text{ skew symmetric} \right\}$$

$$\mathfrak{p} = \left\{ \begin{pmatrix} 0 & B \\ -B^t & 0 \end{pmatrix} \middle| B \text{ an arbitrary } p \times q \text{ matrix} \right\}.$$

The Lie group generated by $\mathfrak{k}$ is $SO(p) \times SO(q)$.

EXERCISES

1. Apply the involution $I_{p,q}$ above to $su(n)$, $n = p + q$ and find $\mathfrak{k}$. The resulting real form is $su(p,q)$.

REPRESENTATION THEORY

CHAPTER 12

Representation Theory

39. The Eight-Fold Way

In this chapter we treat the representations of semi-simple Lie algebras, an important topic for physical applications. We begin with an example to illustrate some of the basic ideas. We discuss representations of the algebra A_2 in the theory of the strong force of nuclear physics. We first provide some background for the physical ideas. For further details the reader may consult Feynman (Book III, Chapter 11-5) or Gibson and Pollard.

When it was observed that protons and neutrons are bound together in the nucleus it was postulated that forces other than electromagnetic forces act between them. This is the so-called 'strong' force. At short range it overpowers the electromagnetic forces by several orders of magnitude. Indeed, electric forces would drive the protons apart (and do under a sufficiently strong perturbation which separates the protons enough to overcome the strong force—this is fission). In 1932 Heisenberg postulated that the proton and neutron are two states of the same particle, called the *nucleon*. Since the strong force is charge independent the strong interactions might be invariant under transformations between linear combinations of the two particle states $|p\rangle$ and $|n\rangle$. The proton and neutron constitute a particle doublet—that is, a family of particles that can be transformed into one another by the operations of the symmetry group. Heisenberg took that symmetry group to be $SU(2)$ and proposed that $|p\rangle$ and $|n\rangle$ transform according to the two dimensional representation of $SU(2)$. In analogy with the theory of electron spin, to which this theory is mathematically equivalent, this group is called *isospin*.

As a basis we take $|p\rangle = \binom{1}{0}$ and $|n\rangle = \binom{0}{1}$. The infinitesimal generators of the representation can be taken to be

$$I_3 = \tfrac{1}{2}\begin{pmatrix} 1 & 0 \\ 0 & -1 \end{pmatrix} \qquad I_+ = \begin{pmatrix} 0 & 1 \\ 0 & 0 \end{pmatrix} \qquad I_- = \begin{pmatrix} 0 & 0 \\ 1 & 0 \end{pmatrix}.$$

We have $I_3|p\rangle = (1/2)|p\rangle$, $I_3|n\rangle = -(1/2)|n\rangle$, $I_+|p\rangle = 0$, $I_-|p\rangle = |n\rangle$, etc.

The electric charge of the particles is obtained as the eigenvalue of the operator $Q = I_3 + (1/2)I = \begin{pmatrix} 1 & 0 \\ 0 & 1 \end{pmatrix}$. The total isospin is given as the eigenvalue of the operator $I^2 = I_1^2 + I_2^2 + I_3^2 = I_+I_- + I_3^2 - I_3$. It is presumed that I_3 and I^2 commute with the Hamiltonian of strong interactions, so that in any interaction involving only the strong force the total isospin and I_3 are conserved. Since I_3 is conserved, so too is the total charge Q. The proton and neutron are particles belonging to a family of particles called *baryons*. Each baryon is assigned a baryon number B, the proton and neutron having baryon number $+1$, and their antiparticles having baryon number -1. There are additional particles which participate in the strong interaction, called mesons, which have baryon number 0. For example, the pions π^+, π^0, and π^- have baryon number 0 and electric charges $1, 0, -1$ respectively. For nucleons and pions the total charge, isospin and baryon number are related by the formula $Q = I_3 + (1/2)B$. Some other mesons and baryons are listed in Figure 12.1.

All these quantities—electric charge, isospin, baryon number, as well as energy and momentum—are conserved in strong interactions; but when the properties of strongly interacting particles were investigated with high energy accelerators it was found that certain interactions which obeyed all these conservation laws nevertheless did not occur. For example (see Feynman) the following reactions fulfill all these conservation laws and yet still do not occur

$$K^- + p \rightarrow K^- + p + K^0$$

$$K^- + p \rightarrow p + \pi^-$$

$$K^- + p \rightarrow \Lambda^0 + K^0.$$

To explain this "strange" behavior Gell-Mann and Nishijima postulated an additional quantity which was to be conserved in any strong interaction, which they called *strangeness*. The relative strangeness of particles could be determined from their behavior in production and decay reactions. The proton and neutron were assigned strangeness zero as a normalizing convention. Below are tables of baryons and mesons with their values of isospin and strangeness. A new quantity, called *hypercharge* and denoted by Y, was introduced. Its relation to the other quantities is given by $Y = S + B$, where S is the strangeness and B is the baryon number. The formula above which related isospin, charge and baryon number was replaced by

$$Q = I_3 + (1/2)Y.$$

This relationship is known as the Gell-Mann–Nishijima formula.

Thus, Heisenberg's original postulate of $SU(2)$ as the symmetry group of strong interactions was not sufficient to explain the observed and non-observed interactions. The additional quantum number Y suggests enlarging the group.

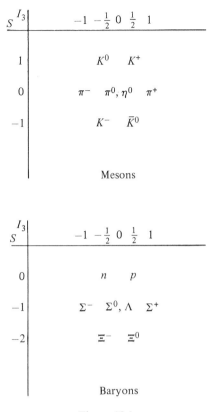

Figure 12.1

Initially an approximation was tried that treated $|p\rangle$, $|n\rangle$, and $|\Lambda^0\rangle$ as equivalent, but this $SU(3)$ theory failed. The theory that worked, now called "flavor $SU(3)$", introduced unobservable particle states, quarks, out of which the observed mesons and baryons could be built. This is the $SU(3)$ model we discuss below as an illustration of representation theory. Today $SU(3)$ is also used in a completely different way as a symmetry called "color $SU(3)$". This is QCD, quantum chromodynamics. An excellent introduction is given in the text by Gottfried and Weisskopf.

In "flavor $SU(3)$" the eight baryons and eight mesons in the figures above each form an eight dimensional multiplet (an octet) that transforms according to the adjoint representation of $su(3)$. The two dimensional Cartan subalgebra gives two quantities which can be simultaneously observed since the corresponding operators commute. These are the isospin I_3 and the hypercharge Y. Each vector of the octet representation represents a particle, and the isospin and hypercharge of that particle are given by the eigenvalues of the corresponding operators I_3 and Y.

The standard three dimensional representation of $SU(3)$ in which a column-vector

$$q = \begin{pmatrix} q_1 \\ q_2 \\ q_3 \end{pmatrix}$$

is transformed by a matrix $U \in SU(3)$ as $q \to Uq$ is denoted by its dimension, 3. A basis is

$$u = \begin{pmatrix} 1 \\ 0 \\ 0 \end{pmatrix} \qquad d = \begin{pmatrix} 0 \\ 1 \\ 0 \end{pmatrix} \qquad \text{and} \quad s = \begin{pmatrix} 0 \\ 0 \\ 1 \end{pmatrix}.$$

These basis vectors are called "quarks". The inequivalent representation in which U is represented by its complex conjugate $\bar{U}$ is denoted by $\bar{3}$. Its basis vectors are called antiquarks and are represented as elements of the dual space by

$$\bar{u} = (1, 0, 0) \qquad \bar{d} = (0, 1, 0) \quad \text{and} \quad \bar{s} = (0, 0, 1).$$

The tensor product of the two representations, $3 \otimes \bar{3}$, is a nine dimensional representation of $SU(3)$. We shall see that $3 \otimes \bar{3}$ decomposes into the direct sum of two irreducible representations, the eight dimensional adjoint representation 8 and a one dimensional identity representation, 1. The vectors in the tensor product space of quarks and anti-quarks thus form an octet of composite particles that transform according to 8, and the mesons are interpreted as the composition of quarks and anti-quarks. The baryons actually come from decomposing the representation $3 \otimes 3 \otimes 3$.

We consider the Lie algebra given by the following matrices:

$$I_3 = (1/2) \begin{pmatrix} 1 & 0 & 0 \\ 0 & -1 & 0 \\ 0 & 0 & 0 \end{pmatrix} \qquad Y = (1/3) \begin{pmatrix} 1 & 0 & 0 \\ 0 & 1 & 0 \\ 0 & 0 & -2 \end{pmatrix} \tag{12.1}$$

$$I_+ = E_{12} = E_\alpha \qquad U_+ = E_{23} = E_\beta \qquad V_+ = E_{13} = E_\gamma$$

$$I_- = (I_+)^t, \qquad U_- = (U_+)^t \quad \text{and} \quad V_- = (V_+)^t.$$

These matrices have the commutation relations

$$[I_3, I_\pm] = \pm I_\pm \qquad\qquad [Y, I_3] = 0$$

$$[I_3, U_\pm] = \mp(\tfrac{1}{2})U_\pm \qquad [Y, U_\pm] = \pm U_\pm \tag{12.2}$$

$$[I_3, V_\pm] = \pm(\tfrac{1}{2})V_\pm \qquad [Y, V_\pm] = \pm V_\pm.$$

Since I_3 and Y commute they are simultaneously diagonalizeable. We represent the simultaneous eigenvectors as states $|\lambda, \mu\rangle$, where λ and μ are the eigenvalues of the operators I_3 and Y on the given eigenvector. The $U_\pm$, $V_\pm$, and $I_\pm$ are shift operators. For example, I_+ shifts the eigenvector $|\lambda, \mu\rangle$ to $|\lambda + 1, \mu\rangle$; that is $I_+|\lambda, \mu\rangle = \text{const.}|\lambda + 1, \mu\rangle$. This follows readily from the

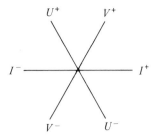

Figure 12.2. Root vectors for A_2.

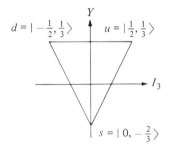

Figure 12.3. Quarks u, d, s; weight diagram for the representation 3.

commutation relations (12.2) for the algebra. From these commutation relations, I_+ raises the isospin (eigenvalue of I_3) by $+1$; U_+ decreases the isospin
by $-\frac{1}{2}$ and raises the hypercharge by $+1$, etc. The action of the shift operators
is thus conveniently represented by the vector figure for A_2 (Fig. 12.2). We
construct what is known as a *weight diagram* for the representation 3. Using
the representation (12.1) we find $I_3 u = \frac{1}{2}u$ and $Yu = \frac{1}{3}u$, so u is represented
by $|\frac{1}{2}, \frac{1}{3}\rangle$, etc. The three vectors u, d, and s are pictured in Figure 12.3.

The *weights* of a representation are functionals on the Cartan subalgebra
whose values are the eigenvalues of the matrices $\rho(I_3)$ and $\rho(Y)$ in the representation. The *weight vectors* are the corresponding eigenvectors. From Figures
12.2 and 12.3 we should have $V_- u = s$; and a simple check shows this is
true. Moreover, $I_- u = d$ and $U_- d = s$; $I_+ s = 0$; etc. The representation $\bar{3}$ is
obtained as follows. Since the antiquarks are row vectors we place them on
the left; for example

$$\bar{U} V_+ = (1,0,0) \begin{pmatrix} 0 & 0 & 1 \\ 0 & 0 & 0 \\ 0 & 0 & 0 \end{pmatrix} = \bar{s}.$$

Now $\rho(A) = -A^t$ is a Lie algebra representation. We obtain the representation of $\bar{3}$ by acting on the right by the negative of the matrices in (12.1). That

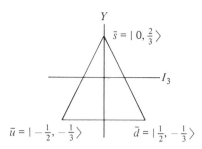

Figure 12.4. Anti-quarks $\bar{u}, \bar{d}, \bar{s}$; weight diagram for $\bar{3}$.

is $\rho(A)\bar{q} = \bar{q}(-A)$ where A is a matrix in the algebra and $\bar{q}$ is an anti-quark. Thus $\rho(I_3)\bar{u} = \bar{u}(-I_3) = -\frac{1}{2}\bar{u}$ and $\rho(V_+)\bar{u} = -\bar{u}V_+ = -\bar{s}$. The weight diagram for the anti-quarks is given in Figure 12.4.

Now let us consider the tensor product representation $3 \otimes \bar{3}$. We first describe the procedure for constructing tensor products of representations. Let V and W be vector spaces which transform under a group $\mathfrak{G}$ according to the representations Γ^1 and Γ^2. We define the tensor product representation $\Gamma = \Gamma^1 \otimes \Gamma^2$ on $V \otimes W$ by $\Gamma(v \otimes w) = (\Gamma^1 v) \otimes (\Gamma^2 w)$ and extend to the entire space $V \otimes W$ by linearity. The infinitesimal generators of this representation act as derivations on the tensor product space. Thus, if L_1 and L_2 are the infinitesimal generators of the representations Γ^1 and Γ^2, we have $L(v \otimes w) = (L_1 v) \otimes w + v \otimes (L_2 w)$. This result is established by the usual argument (see §12). We streamline our notation by dropping the tensor product sign $\otimes$.

Consider the tensor product $u\bar{d}$, for example. We have

$$I_3(u\bar{d}) = (I_3 u)\bar{d} + u(I_3\bar{d})$$
$$= \tfrac{1}{2}u\bar{d} + u(\tfrac{1}{2}\bar{d}) = u\bar{d}$$
$$Y(u\bar{d}) = (Yu)\bar{d} + u(Y\bar{d})$$
$$= \tfrac{1}{3}u\bar{d} + u(-\tfrac{1}{3}\bar{d}) = 0.$$

Therefore the values of the isospin and hypercharge of $u\bar{d}$ are 1 and 0 respectively, and the weights of $u\bar{d}$ are $|1, 0\rangle$. Now the pion π^+ also has these same quantum numbers: $I_3 = 1$ and $Y = 0$; so we identify π^+ with the composite particle $u\bar{d}$. The standard convention is $\pi^+ = -u\bar{d}$.

Now $I_- u\bar{d} = (I_- u)\bar{d} + u(I_-\bar{d}) = d\bar{d} - u\bar{u}$. On the other hand $\{\pi^+, \pi^0, \pi^-\}$ transform as an isospin triplet corresponding to the three dimensional representation D^1 of $SU(2)$. (See the diagram of the mesons in Figure 12.1.) These representations were discussed in Chapter 4. The states are labelled by $|j, m\rangle$ with $m = j, j - 1, \ldots, -j$. In this case, $j = 1$ and $m = 1, 0, -1$. We have

$$\pi^+ = |1, 1\rangle, \qquad \pi^0 = |1, 0\rangle, \qquad \text{and} \quad \pi^- = |1, -1\rangle.$$

One finds (see (4.11)) that $I_-|1, 1\rangle = \sqrt{2}|1, 0\rangle = \sqrt{2}\pi^0$. Thus we conclude

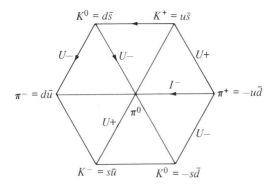

Figure 12.5. Meson octet in $3 \otimes \bar{3}$.

that $\pi^0 = (u\bar{u} - d\bar{d})/\sqrt{2}$. Similarly $I_- \pi^0 = \sqrt{2}\pi^-$ and we obtain $\pi^- = d\bar{u}$. The relationships among the scalar mesons are represented in Figure 12.5.

From the root figure for A_2 we see that $U_+ \pi^+ = \text{const.} K^+$; and since $U_+(-u\bar{d}) = u\bar{s}$, we take $K^+ = u\bar{s}$. Now $U_- K^0 = s\bar{s} - d\bar{d}$; this particle also has zero isospin and hypercharge. In fact, each of the particles $u\bar{u}$, $d\bar{d}$, and $s\bar{s}$ has $I_3 = Y = 0$. The particle

$$\eta' = \frac{u\bar{u} + d\bar{d} + s\bar{s}}{\sqrt{3}}$$

has zero isospin and hypercharge and transforms as a singlet—that is, it is invariant under the group action. (Check that $U_\pm \eta' = V_\pm \eta' = I_\pm \eta' = 0$.) We choose the vector $\eta = \alpha u\bar{u} + \beta d\bar{d} + \gamma s\bar{s}$ to be orthogonal to both η' and π^0, and normalized. We find

$$\eta = \frac{u\bar{u} + d\bar{d} - 2s\bar{s}}{\sqrt{6}}.$$

The pions and kaons together with the particle η transform as an octet—an eight dimensional representation of $SU(3)$. The η' particle transforms as a singlet. The tensor product representation $3 \otimes \bar{3}$ thus decomposes as $8 + 1$. The pions, kaons and η's are mesons. Baryons are viewed as composites of three quarks and thus arise in the representation $3 \otimes 3 \otimes 3 = 10 + 8 + 8 + 1$, where 10 denotes a ten dimensional representation and 8, which occurs twice, is the eight dimensional adjoint representation. The weight diagram for the representation 10 is given in Figure 12.6. The Ω^- particle, which was predicted theoretically by this representation, was discovered in 1964 (cf. Barnes *et al. Phys. Rev. Lett. 12.* (1964), p. 204).

Today several hundred mesons and baryons are known. All are interpreted as bound states of quarks and anti-quarks (mesons) or three quarks (baryons). The number of quark flavors has grown to six (up, down, charmed, strange, top, bottom) and the great variety of mesons and baryons reflects the possible spatial arrangements of these fundamental quarks.

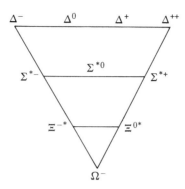

Figure 12.6. Weight diagram for 10.

EXERCISES

1. Decompose $3 \otimes \bar{3}$ into irreducible representations. Show that $3 \otimes \bar{3} = 6 + 3$ where 6 is a six dimensional irreducible representation. Draw the weight diagram.

2. Show $3 \otimes 3 \otimes 3 = 10 + 8 + 8 + 1$ where 10 is a ten dimensional irreducible representation. The eight dimensional representation 8 occurs twice. It is assumed that the baryons $(p, n, \Sigma, \ldots)$ are composed of triplets of quarks.

40. Weights and Weight Vectors

One of the fundamental results of representation theory of semi-simple Lie algebras is that the irreducible representations are uniquely labelled by l-tuples ($l = $ rank $\mathfrak{g}$) of non-negative integers. In particular, the representations may be labelled by placing a non-negative integer at each of the simple roots of the Dynkin diagram. The *fundamental* representations of $\mathfrak{g}$ are those obtained when one of the roots is assigned $+1$ and all the others are assigned 0. For example, the fundamental representations of A_2 are designated by

$$\underset{1}{\alpha_1} \cdot \!\!\!\!\rule[0.35em]{2em}{0.4pt}\!\!\!\! \cdot \underset{0}{\alpha_2} \quad \text{and} \quad \underset{0}{\alpha_1} \cdot \!\!\!\!\rule[0.35em]{2em}{0.4pt}\!\!\!\! \cdot \underset{1}{\alpha_2}.$$

As we shall see, these are, respectively, the representations 3 and $\bar{3}$ discussed in the previous section.

Let ρ be a representation of a semi-simple Lie algebra $\mathfrak{g}$ on a vector space V. For $X \in \mathfrak{g}$ and $v \in V$ we sometimes write Xv for $\rho(X)v$ when the meaning is clear. We shall show below that the operators $\rho(H)$ are diagonalizeable for $H \in \mathfrak{h}$, the Cartan subalgebra of $\mathfrak{g}$. Since the elements of $\mathfrak{h}$ commute, they are simultaneously diagonalizeable. If v is a simultaneous eigenvector then $Hv = \lambda(H)v$ for all $H \in \mathfrak{h}$. The functional λ is linear on $\mathfrak{h}$, and is called a *weight* of the representation; and the subspaces $V_\lambda = \{v | Hv = \lambda(H)v \text{ for all } H \in \mathfrak{h}\}$ are called *weight spaces*. In the case of the adjoint representation the weights were called roots and the weight vectors were called root vectors. The Weyl group $\mathfrak{W}$ acts on the dual space $\mathfrak{h}'$. By exercise 4, §34, the reflection S_α is given by $S_\alpha(\lambda) = \lambda - \lambda(T_\alpha)\alpha$.

Let $\mathfrak{J}$ be the set of all functionals $\{\lambda | \lambda \in \mathfrak{h}', \lambda(T_\alpha) \in \mathbb{Z} \text{ for all } \alpha \in P\}$. Then $\mathfrak{J}$ is a lattice subgroup of $\mathfrak{h}$; that is, if λ_1 and λ_2 belong to $\mathfrak{J}$ then $m\lambda_1 + n\lambda_2 \in \mathfrak{J}$ for all integers m and n. A weight λ is *dominant* if $\lambda(T_\alpha) \geq 0$ for all $\alpha \in P^+$. We construct a basis for $\mathfrak{J}$ as follows. In the Weyl–Chevalley basis the elements T_i and the shift operators E_i corresponding to the positive roots α_i satisfy the commutation relations

$$[E_i, E_{-j}] = \delta_{ij} T_i, \qquad [T_i, E_j] = a_{ij} E_j,$$

where a_{ij} is the matrix of Cartan integers. We choose as a basis for $\mathfrak{J}$ a set of functionals λ_i such that $\lambda_i(T_j) = \delta_{ij}$. Then for $\lambda \in \mathfrak{J}$ we have $\lambda(T_i) = m_i$, and $\lambda = \sum_i m_i \lambda_i$.

Theorem 12.1. *Let ρ be a representation of a semi-simple algebra $\mathfrak{g}$. The weights of ρ are in $\mathfrak{J}$, and the vector space V decomposes into a direct sum of weight spaces: $V = \sum_\lambda V_\lambda$. The set of weights is invariant under the Weyl group $\mathfrak{W}$: If λ is a weight and $\varepsilon = \operatorname{sgn} \lambda(T_\alpha)$ then $\lambda, \lambda - \varepsilon\alpha, \lambda - \varepsilon 2\alpha, \ldots, \lambda - \lambda(T_\alpha)\alpha$ are weights for any $\alpha \in P$. Finally $m_\lambda = \dim V_\lambda = \dim V_{S\lambda}$ for any $S \in \mathfrak{W}$.*

PROOF. Recall that the subalgebras $\{T_\alpha, X_\alpha, X_{-\alpha}\} = A_1$ sit in $\mathfrak{g}$. From the representation theory for A_1 (see Lemma 10.6) the operator $\rho(T_\alpha)$ is diagonalizable; and so V decomposes into a direct sum of eigenvectors of $\rho(T_\alpha)$ for each α. Since the $\{\rho(T_\alpha)\}$ all commute these eigenvectors are simultaneous eigenvectors of the entire family $\{\rho(H) | H \in \mathfrak{h}\}$. Furthermore (Lemma (10.6)), all the eigenvalues of $\rho(T_\alpha)$ in the representations of A_1 were integers. This means all the weights of the representation lie in $\mathfrak{J}$.

Let v_λ be the weight vector corresponding to the weight λ. By the commutation relations $[H, E_\alpha] = \alpha(H)E_\alpha$ it is easy to see that $E_\alpha V_\lambda \subseteq V_{\lambda+\alpha}$. In fact,

$$HE_\alpha v_\lambda = (E_\alpha H + [H, E_\alpha])v_\lambda = (E_\alpha \lambda(H) + \alpha(H)E_\alpha)v_\lambda$$

$$= (\lambda(H) + \alpha(H))v_\lambda.$$

Iterating this result for E_α, we see that $(E_\alpha)^n v_\lambda$ is a weight vector with weight $\lambda + n\alpha$. This result holds for negative n if we define $(E_\alpha)^{-n} = (E_{-\alpha})^n$. Thus we get a string of weights $\lambda + n\alpha$ for some set of integers $p \leq n \leq q$. We leave it as an exercise to show that $(E_{-\alpha})^r v_\lambda \neq 0$ if $r = \lambda(T_\alpha)$. (This follows from the representations for A_1.) Thus $\lambda, \lambda - \alpha, \ldots, \lambda - \lambda(T_\alpha)\alpha$ are weights. It is immediate from this argument that $m_\lambda \leq m_{S\lambda}$; for by applying $(E_{-\alpha})^{\lambda(T_\alpha)}$ to every vector in V_λ we obtain a vector in $V_{S\lambda}$, and these are all linearly independent. Reversing the argument we see that also $m_{S\lambda} \leq m_\lambda$, and hence $m_\lambda = m_{S\lambda}$. This finishes the proof of Theorem 12.1. $\qquad\square$

A weight λ of a representation ρ is said to be a *highest weight* if $\lambda + \alpha$ is not a weight for any positive root α. For example the root $\alpha + \beta$ is the highest weight in the adjoint representation for A_2 (see Figure 10.5). It is clear that every finite dimensional representation has a highest weight; for starting with a weight λ we may add positive roots to λ until we have obtained a highest weight. Given a highest weight λ, let v be the corresponding highest weight vector. We construct the subspace R_λ spanned by the vectors $\{(XY...)v \,|\, X, Y, ... \in \mathfrak{g}\}$. The vector space R_λ is sometimes called a *cyclic* space.

The algebra of matrices generated by all sums and products of the matrices of a representation ρ is called the *enveloping algebra* of the representation, and is denoted by $A_\rho(\mathfrak{g})$. For example, the matrices of the representation D^1 of $su(2)$ are

$$J_3 = \begin{pmatrix} 1 & 0 & 0 \\ 0 & 0 & 0 \\ 0 & 0 & -1 \end{pmatrix} \quad J_+ = \sqrt{2}\begin{pmatrix} 0 & 1 & 0 \\ 0 & 0 & 1 \\ 0 & 0 & 0 \end{pmatrix} \quad J_- = \sqrt{2}\begin{pmatrix} 0 & 0 & 0 \\ 1 & 0 & 0 \\ 0 & 1 & 0 \end{pmatrix}.$$

The enveloping algebra of this representation consists of all sums and products of these 3×3 matrices. The cyclic space R_λ of a representation ρ is obtained as $R_\lambda = A_\rho(\mathfrak{g})v$.

Lemma 12.2. *The vector space R_λ is invariant under ρ; the highest weight is simple; and R_λ is irreducible.*

PROOF. It is clear that R_λ is invariant under the representation ρ. If λ is not simple then there exists a sequence of roots α, β, ... such that $(E_\alpha E_\beta...)v$ also has weight λ. Now $HE_\alpha = E_\alpha(H + \alpha(H))$; and iterating this relationship we get $H(E_\alpha E_\beta...) = (E_\alpha E_\beta...)(H + (\alpha + \beta + \cdots)(H))$. Therefore $(E_\alpha E_\beta...)v$ has weight λ only if $\alpha + \beta + \cdots = 0$. In that case we claim that $(E_\alpha E_\beta...)v$ is a scalar multiple of v. Since $\alpha + \beta + \cdots = 0$ some of the roots must be positive. Commute the shift operators corresponding to the positive roots all the way to the right, where they annihilate v. As they are commuted past the other operators they produce elements in $\mathfrak{h}$. For example, suppose $\alpha + \beta + \gamma = 0$, and that $\alpha \in P^+$ and $\beta, \gamma \in P^-$. Then v is annihilated by E_α, $E_{-\beta}$, and $E_{-\gamma}$, and

$$E_\alpha E_\beta E_\gamma v = (E_\beta E_\alpha + [E_\alpha, E_\beta])E_\gamma v$$

$$= E_\beta(E_\gamma E_\alpha + [E_\alpha, E_\gamma])v + [E_\alpha, E_\beta]E_\gamma v$$

$$= (N_{\alpha\gamma}E_\beta E_{-\beta} + N_{\alpha\beta}E_{-\gamma}E_\gamma)v$$

$$= N_{\alpha\beta}(E_\gamma E_{-\gamma} + [E_{-\gamma}, E_\gamma])v$$

$$= N_{\alpha\beta}\lambda([E_{-\gamma}, E_\gamma])v,$$

since v is a weight vector and $[E_{-\gamma}, E_\gamma] \in \mathfrak{h}$.

To show that R_λ is irreducible it suffices to show that ρ acts transitively on a basis for R_λ. The verification of this step is left to the reader. $\qquad\square$

Theorem 12.3. *An irreducible representation is uniquely determined by its highest weight. Every irreducible representation ρ is isomorphic to the representation of the algebra on a cyclic space R_λ.*

PROOF (Following Samelson). Let ρ and ρ' be two irreducible representations of $\mathfrak{g}$ with highest weight λ; let v and v' be the corresponding highest weight vectors; and let R and R' be the resulting cyclic spaces.

Let $\alpha_1, \ldots, \alpha_l$ be the simple roots of $\mathfrak{g}$ and $E_1, \ldots, E_l$ be the corresponding shift operators. The entire algebra is generated by $E_{\pm 1}, \ldots, E_{\pm l}$ and their commutators. The E_i satisfy the commutation relations $[E_i, E_{-j}] = \delta_{ij} T_i$. (Recall that $[E_\alpha, E_{-\beta}] = 0$ for simple roots α and β, since $\alpha - \beta$ is not a root for simple roots α and β.) Consider a formal infinite dimensional vector space V^λ defined to be that spanned by the basis vectors V_0 and $\{(E_{-i} E_{-j} \ldots) V_0 |$ where $1 \le i, j \ldots \le l\}$. We construct an infinite dimensional representation of $\mathfrak{g}$ on V^λ. First, $T_i V_0 = \lambda(T_i) V_0$, $E_i V_0 = 0$; the action of E_{-k} on V^λ is clear. Thus, V_0 is a highest weight vector for this infinite dimensional representation. Finally, we define the action of E_k by induction.

$$E_k(E_{-i} E_{-j} \ldots) V_0 = (E_{-i} E_k + [E_k, E_{-i}])(E_{-j} \ldots) V_0$$
$$= (E_{-i} E_k + \delta_{ij} T_i)(E_{-j} \ldots) V_0 \qquad (12.3)$$
$$= (E_{-i} E_k \ldots) V_0 + r(T_i)(E_{-j} \ldots) V_0$$

where r is the weight $\lambda - \alpha_j - \cdots$.

We define a map π from V^λ to R as follows:

$$\pi(V_0) = v$$

$$\pi(E_{-i} E_{-j} \ldots) V_0 = (\rho(E_{-i})\rho(E_{-j}) \ldots) v.$$

The mapping π is extended to the whole space by linearity. We claim that π intertwines with the two representations of $\mathfrak{g}$ on V^λ and R. Clearly $\pi(E_{-i} w) = E_{-i}\pi(w)$ for any w in V^λ. As for $\pi(E_i w)$ we again use the commutation relations $[E_i, E_{-j}] = \delta_{ij} T_i$ and proceed by induction, as in (12.3). We thus have that $\pi(Xw) = \rho(X)\pi(w)$ for any $w \in V^\lambda$, and any X in $\mathfrak{g}$. It follows that $\pi(XYw) = \rho(X)\pi(Yw) = \rho(X)\rho(Y)\pi(w)$. Therefore $\ker \pi$ is an invariant subspace of V^λ (under the representation ρ). Finally, it is clear that π maps V^λ onto R.

In the same way we construct a mapping from V^λ to R', denoted by π'. We claim that $\ker \pi = \ker \pi'$. Since $\ker \pi$ is an invariant subspace, $\pi'(\ker \pi)$ is an invariant subspace of R'. (That is, it is invariant under the representation ρ'.) By the irreducibility of R' it is either 0 or R'. But V_0 does not belong to $\ker \pi$, so all vectors in $\ker \pi$ have weight lower than λ. Since π preserves weights all vectors in $\pi'(\ker \pi)$ have weights lower than λ. Therefore v' does not belong to $\pi'(\ker \pi)$ and $\pi'(\ker \pi) = 0$. This shows that $\ker \pi \subseteq \ker \pi'$. Similarly $\ker \pi' \subseteq \ker \pi$; so $\ker \pi = \ker \pi'$.

We next define a linear map φ from R to R' by the same construction. Thus $\varphi(v) = v'$ and $\varphi(\rho(E_{-i})w) = \rho'(E_{-i})\varphi(w)$. One shows, as before, that φ inter-

twines with the representations ρ and ρ': $\varphi\rho(X)\rho(Y) = \rho'(X)\rho'(Y)\varphi$. In particular, φ is a Lie algebra homomorphism from $\rho(\mathfrak{g})$ to $\rho'(\mathfrak{g})$. We must also show
that φ is unambiguously defined. Suppose that $w \in R$ has two distinct representations in terms of the vectors $(E_{-i}E_{-j}\ldots)v$; that is, suppose there are two
vectors W_1 and $W_2 \in V^\lambda$ such that $\pi(W_1) = \pi(W_2) = w$. Then $W_1 - W_2 \in \ker \pi$.
In that case $\pi'(W_1)$ and $\pi'(W_2)$ are both possible values for $\varphi(w)$. But since
$\ker \pi = \ker \pi'$ we see that $\pi'(W_1) = \pi'(W_2)$. To show that φ is an isomorphism,
suppose $\varphi(w) = 0$ for some $w \in R$. Let W be any element in $\pi^{-1}(w)$. Then
$\pi'(W) = \varphi(w)$ and $\pi(W) = w$. If $\varphi(w) = 0$ then $W \in \ker \pi'$, hence $W \in \ker \pi$ and
so $w = \pi(W) = 0$. Hence φ is an isomorphism. $\square$

Let $\{\alpha_1, \ldots \alpha_l\}$ be the fundamental root system for a semi-simple algebra $\mathfrak{g}$
and let $\{T_1, \ldots T_l\}$ be the corresponding operators in the Cartan subalgebra
$\mathfrak{h}$. The Cartan matrix is given by $a_{ij} = \alpha_j(T_i)$. (This matrix can be entirely
determined from the Dynkin diagram.) Let $\lambda \in \mathfrak{I}$, the set of integral forms.
Then $\lambda(T_i)$ is an integer for each of the T_i. We choose a dual basis $\{\lambda_i\}$ such
that $\lambda_i(T_j) = \delta_{ij}$. Setting $\lambda_i = \sum_j M_{ij}\alpha_j$ we obtain

$$\delta_{ij} = \sum_k M_{ik}\alpha_k(T_j) = \sum_k M_{ik}a_{jk}$$

where a_{jk} is the Cartan matrix for $\mathfrak{g}$. Thus $M = (a_{jk}^{-1})^t$. The functionals λ_i are
called the *fundamental* weights, and the irreducible representations associated
with them are called the *fundamental* representations.

For example, let us compute the fundamental weights for the algebra A_2.
The Cartan matrix for A_2 is

$$\begin{pmatrix} 2 & -1 \\ -1 & 2 \end{pmatrix}$$

so we get

$$M(A^t)^{-1} = (\tfrac{1}{3})\begin{pmatrix} 2 & 1 \\ 1 & 2 \end{pmatrix}.$$

For $(1, 0)$ we obtain $\lambda_1 = (2\alpha_1 + \alpha_2)/3$ and for $(0, 1)$ we obtain $\lambda_2 = (\alpha_1 + 2\alpha_2)/3$. We leave it as an exercise to show that the representations corresponding
to these two fundamental weights are the representations which we denoted
by 3 and $\bar{3}$ in §40. We may thus label these by $(1, 0)$ and $(0, 1)$ respectively.

We have seen above that each irreducible representation of a semi-simple
algebra $\mathfrak{g}$ is uniquely determined by its maximal weight. In turn, each maximal
weight λ of a representation is labeled by the non-negative integers $m_i = \lambda(T_i)$.
From our construction of the fundamental weights λ_i this means that $\lambda = \sum_i m_i\lambda_i$.

Theorem 12.4. *For each set of nonnegative integers $m_1, \ldots m_l$ there exists an
irreducible representation of a semisimple algebra $\mathfrak{g}$ of rank l. The highest
weight is given by $\lambda = \sum_i m_i\lambda_i$, where λ_i are the fundamental weights of $\mathfrak{g}$.*

Rather than giving the proof of this theorem, which is somewhat involved, we explicitly construct the fundamental representations of the classical algebras. This will be done in the next section. Cartan himself obtained the result by explicit calculations. Later, a general proof was given by Weyl [1], using the theory of integration on compact groups. An entirely algebraic treatment of the finite dimensional irreducible representations was given by Harish–Chandra. For a proof of this theorem the reader may consult these references, as well as Jacobson, Samelson, or Varadarajan.

We close this section by constructing the fundamental representations 3 and $\bar{3}$ of A_2 using the boson formalism discussed in §16. Let a_i and a_i^* be a set of annihilation and creation operators for $i = 1, 2, 3$. These operators satisfy the commutation relations (cf. (4.7))

$$[a_i, a_j] = [a_i^*, a_j^*] = 0, \quad [a_i, a_j^*] = \delta_{ij} I.$$

We obtain a realization of the algebra A_2 by

$$E_\alpha = a_1^* a_2, \qquad E_\beta = a_2^* a_3, \quad \text{and} \quad E_\gamma = a_1^* a_3$$

$$E_{-\alpha} = a_2^* a_1, \qquad E_{-\beta} = a_3^* a_2, \quad \text{and} \quad E_{-\gamma} = a_3^* a_1$$

$$T_1 = a_1^* a_1 - a_2^* a_2, \qquad T_2 = a_2^* a_2 - a_3^* a_3.$$

$$I_3 = (\tfrac{1}{2}) T_1, \qquad Y = (\tfrac{1}{3})(T_1 + 2T_2).$$

The vacuum state is denoted by $|0\rangle$. The three "quarks" are given by $q_i = a_i^* |0\rangle$. One easily finds from the commutation relations for the a_i, a_i^* that the vectors q_1, q_2, and q_3 transform as the quarks u, d, and s in the representation 3 discussed in §39. For example, $E_{-\alpha} q_1 = a_2^* a_1 q_1 = a_2^* |0\rangle = q_2$, etc.

The representation $\bar{3}$ is obtained by introducing a *second* set of creation and annihilation operators b_i, b_i^* for $i = 1, 2, 3$ and setting $E_\alpha = -b_2^* b_1$, $E_\beta = -b_3^* b_2$, and $E_\gamma = -b_3^* b_1$, etc. The antiquarks are $\bar{q}_i = b_i^* |0\rangle$, with $\bar{q}_3 = \bar{s}$, $\bar{q}_2 = \bar{d}$, and $\bar{q}_1 = \bar{u}$. Now, for example, we get $E_\alpha \bar{q}_1 = -\bar{q}_2$, etc.

Exercises

1. Show that $(1, 0)$ and $(0, 1)$ are the representations 3 and $\bar{3}$ discussed in §40. (*Hint:* compare the highest weights.)

41. Tensor Products

Let φ be a representation of a Lie algebra on a vector space V. We discuss the representations induced by φ on the tensor products of V. If $\{v_1, v_2, \ldots v_n\}$ is a basis for V, the tensor product $V \otimes V$ is spanned by the tensor products $\{v_i \otimes v_j | 1 \leq i, j \leq n\}$. It is a space of dimension n^2. The representation induced by φ on $V \otimes V$, denoted by $\varphi \otimes \varphi$, is given by

$$\varphi \otimes \varphi(v_i \otimes v_j) = (\varphi v_i) \otimes v_j + v_i \otimes (\varphi v_j).$$

As we have said before, this is a result of the fact that Lie derivatives act as derivations on tensor products; and the representation φ acting on V must be considered as the Lie derivative of the corresponding group acting on V.

It is an immediate consequence of this action that the weights are additive. That is, if v and w are weight vectors with weights λ and μ, then $v \otimes w$ is a weight vector of $\varphi \otimes \varphi$ with weight $\lambda + \mu$. Therefore, if $\{\lambda_i\}$ is a complete set of weights for φ then $\{\lambda_i + \lambda_j\}$ is a complete set of weights for $\varphi \otimes \varphi$. Furthermore, if v is the highest weight vector for φ then $v \otimes v$ will be the highest weight vector for $\varphi \otimes \varphi$.

The symmetric group S_2 acts naturally on $V \otimes V$ by simply interchanging the two factors: if σ denotes the transposition (12) then $\sigma(v \otimes w) = w \otimes v$. The vector space $V \otimes V$ decomposes into two invariant subspaces under σ: the even and the odd tensor products:

$$V \otimes_s V = \{v_i \otimes v_j + v_j \otimes v_i | 1 \leq i, j \leq n\}$$

$$V \wedge V = \{v_i \otimes v_j - v_j \otimes v_i | 1 \leq i < j \leq n\}.$$

The space $V \otimes_s V$ is the space of symmetric tensors; it has dimension $n(n + 1)/2$ and σ has eigenvalue $+1$ on it. The space $V \wedge V$ is the space of skew symmetric tensors; it has dimension $n(n - 1)/2$ and σ has eigenvalue -1 on it.

It is clear that σ commutes with φ, and therefore φ leaves $V \otimes_s V$ and $V \wedge V$ invariant. Thus the representation $\varphi \otimes \varphi$ reduces to a sum of representations on the two subspaces of symmetry classes of tensors. We denote these respectively by $\varphi \otimes_s \varphi$ and $\varphi \wedge \varphi$. For example, in the case of $su(3)$ we have the decomposition $3 \otimes 3 = 6 + 3$. The six dimensional representation is the action on the symmetric tensor products, while the three dimensional representation is the action on the antisymmetric tensor products.

Let us consider the tensor product $D^{1/2} \otimes D^{1/2}$ where $D^{1/2}$ is the fundamental representation of $su(2)$. The computation of tensor products of representations of $su(2)$ is known in physics as the "addition of angular momenta." The tensor product $D^{1/2} \otimes D^{1/2}$ represents the coupling of two particles of spin $\frac{1}{2}$ and its decomposition gives the possible spin states of the coupled system. Denote the two states for the representation $D^{1/2}$ by u and d (for spin "up" and spin "down"). We have

$$J_+ u = 0 \qquad J_- u = d \qquad J_+ d = u \qquad J_- d = 0$$

$$J_3 u = \tfrac{1}{2} u \qquad J_3 d = -\tfrac{1}{2} d.$$

The tensor products are

$$V \otimes_s V = \{u \otimes u, \, u \otimes d + d \otimes u, \, d \otimes d\}$$

$$V \wedge V = \{u \otimes d - d \otimes u\}.$$

The tensor product $D^{1/2} \otimes D^{1/2}$ thus decomposes into the sum of a three dimensional representation and a one dimensional representation. Let us look at the antisymmetric "singlet": $u \otimes d - d \otimes u = 2u \wedge d$. It has weight

$\frac{1}{2} + (-\frac{1}{2}) = 0$. Furthermore, $J_+(u \wedge d) = (J_+u) \wedge d + u \wedge (J_+d) = u \wedge u = 0$. Similarly $J_-u \wedge d = 0$. The singlet state is thus annihilated by all operators in the algebra and is therefore invariant under the group action. The corresponding representation, denoted by D^0, is therefore the identity representation.

The highest weight vector for the symmetric tensors, $u \otimes u$, has weight 1. Moreover, $J_+u \otimes u = 0$; $J_-u \otimes u = d \otimes u + u \otimes d$; and $J_-(d \otimes u + u \otimes d) = d \otimes d$. The triplet thus transforms according to the representation D^1 of $su(2)$. We write this decomposition as $D^{1/2} \otimes D^{1/2} = D^1 + D^0$. More generally, the following decomposition, known as the *Clebsch–Gordon* series, is true for the representations D^j of $su(2)$:

$$D^j \otimes D^k = D^{j+k} + D^{j+k-1} + \cdots + D^{|j-k|}.$$

(Cf. Wigner, [2].)

Let us return to the fundamental representations for A_2 and show that the representation $\bar{3}$ can be obtained as $3 \wedge 3$, that is, as the restriction of the tensor product representation $3 \otimes 3$ to the antisymmetric tensor products. This is an important construction which can be used to obtain most of the fundamental representations for the other classical algebras.

Let us consider the three quarks in the representation 3. Taking their wedge products we have

$$q_2 \wedge q_3 = \frac{q_2 \otimes q_3 - q_3 \otimes q_2}{2}$$

$$q_3 \wedge q_1 = \frac{q_3 \otimes q_1 - q_1 \otimes q_3}{2}$$

$$q_1 \wedge q_2 = \frac{q_1 \otimes q_2 - q_2 \otimes q_1}{2}.$$

It is easily seen that the three vectors $q_2 \wedge q_3, q_3 \wedge q_1$, and $q_1 \wedge q_2$ transform like the antiquarks $\bar{q}_1$, $\bar{q}_2$, and $\bar{q}_3$. For example, we have $E_\alpha q_2 \wedge q_3 = a_1^* a_2(q_2 \wedge q_3) = (a_1^* a_2)q_2 \wedge q_3 + q_2 \wedge (a_1^* a_2)q_3 = q_1 \wedge q_3 = -q_3 \wedge q_1 = -\bar{q}_2$. On the other hand, for the antiquarks, $E_\alpha = -b_2^* b_1$, so $E_\alpha \bar{q}_1 = -\bar{q}_2$. Similarly, the action of the two representations $3 \wedge 3$ and $\bar{3}$ agree on the entire space.

The above construction can be carried through quite generally to obtain some (but not all) of the fundamental representations of the classical algebras as anti-symmetric tensor products of the standard representation. In the case of A_l we can get all the fundamental representations by this process. The operators T_i are given by

$$T_1 = \text{diag}(1, -1, 0, \ldots), \qquad T_2 = \text{diag}(0, 1, -1, 0 \ldots),$$

$$\ldots T_l = \text{diag}(0, \ldots 1, -1).$$

The fundamental weights are given by $\lambda_i(T_j) = \delta_{ij}$. For $H = \text{diag}(h_{11}, h_{22}, \ldots)$ we define $\omega_i(H) = h_{ii}$. Then it is easily seen that

$$\lambda_1 = \omega_1, \qquad \lambda_2 = \omega_1 + \omega_2, \qquad \lambda_3 = \omega_1 + \omega_2 + \omega_3, \ldots .$$

The representation corresponding to the weight λ_1 is the standard representation of A_l as the set of traceless $(l+1) \times (l+1)$ matrices. The weight vectors are $e_j = \text{col}(0, \ldots, 1, 0 \ldots)$, etc; e_j has weight ω_j. Recall that the roots of A_l are $\omega_i - \omega_j$ and that the positive simple roots are $\omega_i - \omega_{i+1}$. We have $\omega_2 = \omega_1 - (\omega_1 - \omega_2)$, $\omega_3 = \omega_1 - (\omega_1 - \omega_2) - (\omega_2 - \omega_3)$, etc. Thus all the other weights ω_i are obtained by subtracting positive simple roots from ω_1. This shows that ω_1 is the highest weight.

Denote this representation by φ and let $\varphi_2 = \varphi \wedge \varphi$. The basis for $V \wedge V$ is $\{e_i \wedge e_j, 1 \le i < j \le l+1\}$ and $e_i \wedge e_j$ is a weight vector with weight $\omega_i + \omega_j$. In particular, $e_1 \wedge e_2$ has weight $\omega_1 + \omega_2$; it is the highest weight vector for the representation φ_2. Similarly the highest weight for the representation $\varphi_3 = \varphi \wedge \varphi \wedge \varphi$ is $\omega_1 + \omega_2 + \omega_3$, the corresponding weight vector being $e_1 \wedge e_2 \wedge e_3$. In this way we get all the fundamental representations for A_l.

Now let us discuss the fundamental representations of C_l. Let the simple roots be $\omega_1, \ldots \omega_l$. The corresponding elements of $\mathfrak{h}$ are

$$T_1 = \text{diag}(a_0, -a_0, 0, \ldots), \qquad T_2 = \text{diag}(0, a_0, -a_0, \ldots),$$

$$\ldots T_{l-1} = \text{diag}(0 \ldots a_0, -a_0)$$

$$T_l = \text{diag}(0 \ldots a_0).$$

Recall (cf. (10.6)) that the matrices representing $sp(2l, C)$ are $l \times l$ supermatrices whose entries are 2×2 matrices. It is easily seen that the fundamental weights are again $\lambda_j = \alpha_1 + \cdots \alpha_j$; so in the case of C_l as in the case of A_l, all the fundamental representations are obtained as wedge products of the standard one.

The situation for the algebras B_l is different. Here the basic elements of $\mathfrak{h}$ are found to be

$$T_1 = \text{diag}(a_0, -a_0, 0, \ldots), \ldots, \qquad T_{l-1} = \text{diag}(\ldots, a_0, -a_0),$$

$$T_l = 2 \, \text{diag}(0, \ldots, a_0).$$

In this case the fundamental weights are (exercise 2, p. 151)

$$\lambda_1 = \alpha_1, \ldots \qquad \lambda_2 = \alpha_1 + \alpha_2, \ldots$$

$$\lambda_{l-1} = \alpha_1 + \alpha_2 + \cdots + \alpha_{l-1}$$

and

$$\lambda_l = (\tfrac{1}{2})(\alpha_1 + \cdots \alpha_l).$$

The fundamental representations corresponding to the first $l - 1$ weights are obtained as wedge products of the standard representation, as before; but the representation corresponding to λ_l, denoted by Δ_l, cannot be obtained in this way. It is called a *spinor* representation of B_l and will be constructed in Chapter 13. The spinor representations of the orthogonal groups were first obtained by Brauer and Weyl.

Finally, in the case of D_l one finds, by similar computations, that

$$\lambda_1 = \alpha_1, \qquad \lambda_2 = \alpha_1 + \alpha_2, \qquad \ldots \qquad \lambda_{l-2} = \alpha_1 + \cdots \alpha_{l-2},$$

$$\lambda_{l-1} = (\tfrac{1}{2})(\alpha_1 + \cdots + \alpha_{l-1}), \qquad \lambda_l = (\tfrac{1}{2})(\alpha_1 + \cdots \alpha_l).$$

In this case there are two spinor representations corresponding to λ_{l-1} and λ_l, denoted by Δ_l^- and Δ_l^+.

The situation for higher order tensor products is complicated by the fact that there are more symmetry classes of tensors to deal with. For an nth order tensor product, denoted by $V^{\otimes n}$ we must consider the action of S_n on the tensor products. The representations of semi-simple Lie algebras is intertwined with the representation theory of the symmetric group in a complicated way. See Miller, Weyl [2], or Hamermesh.

EXERCISES

1. Obtain the decomposition $D^1 \otimes D^1 = D^2 + D^1 + D^0$ for the standard representation of $so(3)$ by the Lie algebra methods of this section. Show it is equivalent to the action OAO^t, where $O \in SO(3)$ and A is a 3×3 matrix.

2. Obtain the decomposition $3 \otimes 3 = 6 + \bar{3}$ for the Lie algebra $su(3)$.

42. Enveloping Algebras and Casimir Operators

Let ρ be a representation of a semi-simple Lie algebra $\mathfrak{g}$. We define a trace form for ρ by $K_\rho(X, Y) = \operatorname{Tr} \rho(X)\rho(Y)$. The metric tensor for the representation ρ is then $g_{ij} = K_\rho(X_i, X_j)$, where X_i is a basis for $\mathfrak{g}$.

Lemma 12.5. *The trace form of a faithful representation of a semi-simple Lie algebra is non-degenerate.*
(A faithful representation is one with a trivial kernel.)

PROOF. This is, of course, a generalization of Cartan's criterion, Theorem 9.2, to representations other than the adjoint representation. Let $\operatorname{rad} K_\rho = \{X | K_\rho(X, Y) = 0 \text{ for all } Y \in \mathfrak{g}\}$. From the identity $K_\rho([X, Y], Z) + K_\rho(Y, [X, Z]) = 0$ we see that $\operatorname{rad} K_\rho$ is an ideal. By Theorem 9.15 $\operatorname{rad} K_\rho$ is solvable, and therefore a solvable subalgebra of the matrix algebra $\{\rho(X) | X \in \mathfrak{g}.\}$ Since the representation is faithful, the matrix algebra $\rho(\mathfrak{g})$ is isomorphic to $\mathfrak{g}$ and therefore itself semi-simple. Hence $\operatorname{rad} K_\rho$ must be trivial. $\square$

The *Casimir* operator for the representation ρ is the operator

$$C_\rho = g^{ij} X_i X_j.$$

g^{ij} is the inverse of the matrix g_{ij}; we have dropped the ρ and simply written

X_i for the matrix $\rho(X_i)$. The Casimir operator lies in the enveloping algebra of the representation. Setting $X^i = g^{ij}X_j$ we may write C_ρ in the form $C_\rho = X_iX^i$; the matrices X^i are called the dual basis.

For example, the matrices of the representation D^1 of $su(2)$ were given above. The metric tensor for this representation is

$$g = \begin{pmatrix} 2 & 0 & 0 \\ 0 & 0 & 4 \\ 0 & 4 & 0 \end{pmatrix}$$

and the matrices of the dual basis are given by $J^3 = (1/2)J_3$; $J^+ = (1/4)J_-$; $J^- = (1/4)J_+$. The Casimir operator for this representation is $C_\rho = I$.

In fact, whenever ρ is irreducible the Casimir operator will be a scalar multiple of the identity. This follows from *Schur's lemma: Let A be a set of operators on a complex vector space V and suppose that V is irreducible under A, that is, that no proper subspace of V is invariant under the entire set of operators in A. If C commutes with all operators in A then C is a scalar multiple of the identity.*

Let us show that the Casimir operator lies in the center of the enveloping algebra of the representation, that is, that it commutes with all X in the representation. First, for any X we have

$$[X, X_i] = a_{ij}(X)X_j \qquad [X, X^i] = b_{ij}(X)X^j.$$

So

$$a_{ij}(X) + b_{ji}(X) = \text{Tr}[X, X_i]X^j + \text{Tr}[X, X^j]X_i = 0.$$

Hence

$$[X, C] = [X, X_iX^i] = [X, X_i]X^i + X_i[X, X^i]$$
$$= a_{ij}(X)X_jX^i + b_{ij}(X)X_iX^j = 0.$$

EXERCISES

1. Let λ be a weight and α a root. Show that the "α string of weights through λ" is an uninterrupted string of weights $\lambda + n\alpha$, with $p \le n \le q$, where $p \le 0, q \ge 0$, and

$$\frac{2\langle \lambda, \alpha \rangle}{\langle \alpha, \alpha \rangle} = \lambda(T_\alpha) = -(p + q).$$

2. Prove that $\text{Tr}\, C_\rho = \dim \rho$, the dimension of the representation ρ.

3. Find the Casimir operator for the representation D^j of $su(2)$ given by (4.11).

4. Find the highest weight vector for the standard representations of B_l, C_l, and D_l.

CHAPTER 13

Spinor Representations

43. Looking Glass Zoo

Here is a problem whose solution involves a clever application of the Dirac matrices. It was proposed by Sir Arthur Eddington and appeared in *The New Statesman and Nation*. December 19, 1936 (p. 1044).

I took some nephews and nieces to the Zoo, and we halted at a cage marked

> Tovus Slithius, male and female.
> Beregovus Mimsius, male and female.
> Rathus Momus, male and female.
> Jabberwockius Vulgaris, male and female.

The eight animals were asleep in a row, and the children began to guess which was which. "That one at the end is Mr. Tove." No, no! It's Mrs. Jabberwock," and so on. I suggested that they should each write down the names in order from left to right, and offered a prize to the one who got most names right.

As the four species were easily distinguished, no mistake would arise in pairing the animals; naturally a child who identified one animal as Mr. Tove identified the other animal of the same species as Mrs. Tove.

The keeper, who consented to judge the lists, scrutinised them carefully. "Here's a queer thing. I take two of the lists, say, John's and Mary's. The animal which John supposes to be the animal which Mary supposes to be Mr. Tove is the animal which Mary supposes to be the animal which John supposes to be Mrs. Tove. It is just the same for every pair of lists, and for all four species.

"Curiouser and curiouser! Each boy supposes Mr. Tove to be the animal which he supposes to be Mr. Tove; but each girl supposes Mr. Tove to be the animal which she supposes to be Mrs. Tove. And similarly for the other animals. I mean, for instance, that the animal Mary calls Mr. Tove is really Mrs. Rathe, but the animal she calls Mrs. Rathe is really Mrs. Tove."

"It seems a little involved," I said, "but I suppose it is a remarkable coincidence."

"Very remarkable," replied Mr. Dodgson (whom I had supposed to be the keeper) "and it could not have happened if you had brought any more children."
How many nephews and nieces were there? Was the winner a boy or a girl? And how many names did the winner get right?

You may check your solution against the published solution in the January 9, 1937 issue of *The New Statesman and Nation*.

44. Clifford Algebras

In his search for a relativistic theory of the electron, Dirac was led to factor the Klein–Gordon equation

$$U_{tt} - \Delta U - m^2 U = 0$$

into a first order system of differential equations. To this end he proposed finding an operator of the form $\gamma^\mu \partial_\mu$ which would be Lorentz covariant and such that

$$(\gamma^\mu \partial_\mu)^2 = \partial_t^2 - \Delta.$$

Formally squaring this operator one is led to the conditions on the γ's:

$$\{\gamma^\mu, \gamma^\nu\} = 2g^{\mu\nu}I, \qquad \mu, \nu = 0, \dots 3$$

where $g^{\mu\nu}$ is the Minkowski metric tensor introduced in Chapter 1, and $\{\ ,\ \}$ is the anticommutator: $\{A, B\} = AB + BA$. Dirac found a set of 4×4 matrices which satisfied these conditions and thereby constructed his relativistic equation for the electron. The resulting equation, known as the Dirac equation, ultimately led him to predict the existence of anti-matter Dirac [2].

Replacing γ^0 by $i\alpha^0$ and γ^i by α^i we find a set of conditions for the α's, namely

$$\{\alpha^\mu, \alpha^\nu\} = 2\delta_{\mu\nu}. \tag{13.1}$$

The algebra generated by a set of quantities satisfying these relations is called a *Clifford Algebra*. For example, if we wanted to write the Laplacian as a square of an operator $\sum_i \alpha_i \partial/\partial x_i$ we would arrive at the equations (13.1) for the α_i. Below we shall explain how to construct a set of matrices satisfying these anticommutation relations. Once the α's are obtained we construct their products, *viz.*

$$1, \qquad \alpha_i, \qquad \alpha_i \alpha_j, \dots \alpha_1 \alpha_2 \dots \alpha_n. \tag{13.2}$$

Since the α's anticommute (that is, $\alpha_2 \alpha_1 = -\alpha_1 \alpha_2$), we can always arrange their products so that the factors appear in increasing order. The vector space thus generated contains 2^n elements, assuming i runs from 1 to n.

For example, for $n = 2$ we have

$$\alpha_1 = \begin{pmatrix} 0 & 1 \\ 1 & 0 \end{pmatrix} \qquad \alpha_2 = \begin{pmatrix} 0 & i \\ -i & 0 \end{pmatrix}.$$

Then $\alpha_1^2 = \alpha_2^2 = 1$ and $\{\alpha_1, \alpha_2\} = 0$.

By exercise 1, §4, the Pauli spin matrices satisfy $\{\sigma_i, \sigma_j\} = 2\delta_{ij}$ for $i, j = 1$, 2, 3. It is a simple calculation to show that the matrices given by the super-matrices

$$\gamma_1 = \begin{pmatrix} \sigma_1 & 0 \\ 0 & \sigma_1 \end{pmatrix} \qquad \gamma_2 = \begin{pmatrix} \sigma_2 & 0 \\ 0 & \sigma_2 \end{pmatrix} \qquad \gamma_3 = \begin{pmatrix} 0 & \sigma_3 \\ \sigma_3 & 0 \end{pmatrix}$$

$$\gamma_0 = \begin{pmatrix} 0 & -i\sigma_3 \\ i\sigma_3 & 0 \end{pmatrix}$$

constitutes a solution to the problem for $n = 4$.

A general solution of (13.1) is obtained by taking tensor products of the basic ones as follows. We consider a vector space of elements of the form $\psi = (\psi_1, \dots \psi_n)$ where ψ_i is an element of the two dimensional vector space C^2. The ψ's are called spinors and may be regarded as elements of $(C^2)^{\otimes n}$. We construct linear operators on this space of the form $A = A_1 \otimes \cdots \otimes A_n$ where A_i is a 2×2 matrix. The action of A on ψ is given by $A\psi = A_1\psi_1 \otimes \cdots \otimes A_n\psi_n$. Moreover $A \otimes B = A_1 B_1 \otimes \cdots \otimes A_n B_n$. A general solution of the Clifford algebra for even $n = 2m$ is obtained by setting

$$\alpha_j = \sigma_3 \otimes \sigma_3 \otimes \dots \sigma_3 \otimes \overset{j}{\sigma_1} \otimes 1 \dots \otimes 1 \qquad j \le m$$

$$\alpha_{j+m} = \sigma_3 \otimes \dots \sigma_3 \otimes \sigma_2 \otimes 1 \dots \otimes 1.$$

For example, when $m = 2$ we have

$$\alpha_1 = \sigma_1 \otimes 1 \qquad \alpha_2 = \sigma_3 \otimes \sigma_1 \qquad \alpha_3 = \sigma_2 \otimes 1 \qquad \alpha_4 = \sigma_3 \otimes \sigma_2.$$

Then $\{\alpha_1, \alpha_2\} = \sigma_1\sigma_3 \otimes \sigma_1 + \sigma_3\sigma_1 \otimes \sigma_1 = \{\sigma_1, \sigma_3\} \otimes \sigma_1 = 0$, and so forth. For odd $n = 2m + 1$ the matrix α_{2m+1} is given by

$$\alpha_{2m+1} = \sigma_3 \otimes \cdots \otimes \sigma_3.$$

We are going to use the matrices in the Clifford algebra (13.1) to construct the spinor representations of the orthogonal groups which were discussed in the previous chapter. We begin by proving:

Theorem 13.1. Let α_i satisfy the anticommutation relations (13.1) for $i = 1, \dots n$. Then the elements $M_{ij} = \frac{1}{2}\alpha_i\alpha_j$ form the Lie algebra $so(n)$.

PROOF. This theorem is proved by a straightforward calculation of the com-mutators. We have

$$[M_{ij}, M_{rs}] = (\tfrac{1}{4})(\alpha_i\alpha_j\alpha_r\alpha_s - \alpha_r\alpha_s\alpha_i\alpha_j).$$

Now

$$\alpha_r\alpha_s\alpha_i\alpha_j = \alpha_r(2\delta_{is} - \alpha_i\alpha_s)\alpha_j = \cdots$$

$$= 2(\delta_{is}\alpha_r\alpha_j - \delta_{ir}\alpha_s\alpha_j + \delta_{sj}\alpha_i\alpha_r - 2\delta_{rj}\alpha_i\alpha_s) + \alpha_i\alpha_j\alpha_r\alpha_s;$$

so

$$[M_{ij}, M_{rs}] = \delta_{is}M_{jr} + \delta_{ir}M_{sj} + \delta_{sj}M_{ri} + \delta_{rj}M_{is}.$$

These are precisely the commutation relations for the matrices $M_{ij} = E_{ij} - E_{ji}$ which generate the orthogonal groups. □

Consider the vector space V spanned by $\alpha_1, \ldots \alpha_n$, and write $X = \sum_i x^i \alpha_i$. From the commutation relations for the α's we find

$$S_{ij}(\theta) = \exp\{-(\theta/2)\alpha_i\alpha_j\} = \cos(\theta/2)I - \alpha_i\alpha_j \sin(\theta/2).$$

The action $S_{ij}(\theta) X S_{ij}(-\theta)$ is easily seen to be a rotation in V. For example,

$$S_{ij}(\theta)\alpha_i S_{ij}(-\theta) = \alpha_i \cos\theta + \alpha_j \sin\theta$$

$$S_{ij}(\theta)\alpha_j S_{ij}(-\theta) = -\alpha_i \sin\theta + \alpha_j \cos\theta;$$

and all the other α's remain fixed. Thus the action is indeed a rotation, and we have obtained a representation of $SO(n)$.

Now, however, consider the 2^n dimensional representation $\exp\{-(\theta/2)\alpha_i\alpha_j\}$ acting on the vector space spanned by the products (13.2). We are going to show that its infinitesimal generator is the spinor representation of the Lie algebra $so(n)$.

Recall (§36) that it was not convenient to work in the standard basis for the algebras $so(n)$. In the case of even $n = 2l$ we made a unitary transformation in R^n to new coordinates in which the metric tensor was transformed into

$$\begin{pmatrix} 0 & 1 & & \\ 1 & 0 & & \\ \hline & & 0 & 1 \\ & & 1 & 0 \\ & & & & \ddots \end{pmatrix}.$$

We make the corresponding transformation of the Clifford algebra. That is, we set

$$\sigma_1 = \frac{\alpha_1 + i\alpha_2}{\sqrt{2}} \qquad \sigma_1^* = \frac{\alpha_1 - i\alpha_2}{\sqrt{2}}, \qquad \sigma_2 = \frac{\alpha_3 + i\alpha_4}{\sqrt{2}} \quad \text{etc.}$$

We make one more transformation: we set $a_i = \sigma_i/\sqrt{2}$ and $a_i^* = \sigma_i^*/\sqrt{2}$. When this is done we find that the a's satisfy the anticommutation relations

$$\{a_i, a_j\} = 0 \qquad \{a_i^*, a_j^*\} = 0 \qquad \{a_i, a_j^*\} = \delta_{ij}\mathbb{1}. \tag{13.3}$$

The anticommutation relations (13.3) are the relations satisfied by Fermion creation and annihilation operators in quantum mechanics. These are operators associated with particles of half integer spin. We see that they arise mathematically as an artifact of our unitary transformation of R^n. When n is odd we have an additional operator α_0 which we set equal to σ_0.

Let us first consider the case where n is even, $n = 2l$. It is easily seen that the operators

$$
\begin{array}{lll}
a_i a_j & i < j & -(\omega_i + \omega_j) \\
a_i^* a_j^* & i < j & (\omega_i + \omega_j) \\
a_i^* a_j & i \neq j & (\omega_i - \omega_j) \\
a_i^* a_i - \frac{1}{2} & i = j &
\end{array}
\tag{13.4}
$$

generate a Lie algebra isomorphic to the real form (10.7) of D_l given in §36. A general element H in the Cartan subalgebra is given by

$$
H = \sum_i h_i(a_i^* a_i - \tfrac{1}{2});
$$

and the root vectors are the elements $a_i a_j$, $a_i^* a_j^*$, and $a_i a_j^*$ for $i \neq j$. These have roots $-(\omega_i + \omega_j)$, $(\omega_i + \omega_j)$, and $(\omega_j - \omega_i)$ respectively. (We define the functional ω_i by $\omega_i(H) = h_i$.)

Recall that the spinor representations of D_l had highest weights

$$
\lambda_{l-1} = \tfrac{1}{2}(\omega_1 + \omega_2 + \cdots - \omega_l)
$$

and

$$
\lambda_l = \tfrac{1}{2}(\omega_1 + \omega_2 + \cdots + \omega_l).
$$

We obtain a representation with highest weight λ_l as follows. Let $|0\rangle$ be the vacuum state and consider the vector $a_1^* a_2^* \ldots a_l^* |0\rangle$. It is easily seen that this vector has weight λ_l. Moreover, it is annihilated by all the root vectors with positive roots, viz. $a_i^* a_j^*$, and $a_i a_j^*$. It therefore serves as the highest weight vector in a representation of the given Lie algebra.

We obtain the representation with weight λ_{l-1} as follows. The transposition $a_l \leftrightarrow a_l^*$, with all other elements fixed preserves the relations (13.3) and so is an automorphism of the Clifford algebra. On the other hand, given any set of matrices satisfying (13.3), the matrices in (13.4) satisfy the commutation relations of (10.7) and so give a real form of $so(2n, C)$. The new one is obtained from the old one by an automorphism. For example, $a_i a_l^* \to a_i a_l$. But now

$$
H = \sum_{j=1}^{l-1} h_j(a_j^* a_j - \tfrac{1}{2}) + h_l(a_l a_l^* - \tfrac{1}{2})
$$

$$
= \sum_{j=1}^{l-1} h_j(a_j^* a_j - \tfrac{1}{2}) - h_l(a_l^* a_l - \tfrac{1}{2});
$$

and the weight of the vector $a_1^* \ldots a_l^* |0\rangle$ is λ_{l-1}.

A realization of the algebra B_l by Fermion operators is obtained in the same way. Setting

$$
a_0 = \frac{\sigma_0}{2}, \qquad a_1 = \frac{\sigma_1 + i\sigma_2}{2} \qquad a_1^* = \frac{\sigma_1 - i\sigma_2}{2}, \qquad \text{etc.,}
$$

we obtain an algebra isomorphic to the real form (10.8) of B_l as

$$
\begin{array}{lll}
a_0 a_j & -\omega_j & \\
a_0^* a_j^* & \omega_j & \\
a_i a_j & -(\omega_i + \omega_j) & i < j \\
a_i^* a_j^* & (\omega_i + \omega_j) & i < j \\
a_i a_j^* & \omega_j - \omega_i & i \neq j \\
a_i^* a_i - \tfrac{1}{2} & & 1 \leq i \leq l
\end{array}
\tag{13.5}
$$

Again, a general element of the Cartan subalgebra is an operator of the form $H = \sum_j h_j(a_j^* a_j - \tfrac{1}{2})$. As in the case of $so(2l)$ the highest weight vector for the spinor representation is $a_1^* a_2^* \ldots a_l^* |0\rangle$; it has weight $\tfrac{1}{2}(\alpha_1 + \ldots \alpha_l)$.

We close with a realization of the symplectic algebras $sp(2l)$ by boson operators. Let a_i, a_i^* satisfy the commutation relations for boson creation and annihilation operators given in (4.7), viz.

$$
[a_i, a_j] = [a_i^*, a_j^*] = 0, \qquad [a_i, a_j^*] = \delta_{ij} 1.
$$

The root vectors of the algebra (10.6) and their roots are

$$
\begin{array}{lll}
a_i a_j & -(\omega_i + \omega_j) & i < j \\
a_i^* a_j^* & (\omega_i + \omega_j) & i < j \\
a_i^* a_j & (\omega_i - \omega_j) & i \neq j \\
a_i^2 & -2\omega_i & \\
(a_i^*)^2 & 2\omega_i. &
\end{array}
\tag{13.6}
$$

One important application of spinor representations to a problem in mathematical physics is in the Onsager theory of phase transitions in the Ising model. Onsager's original evaluation of the partition function employed Lie algebras. His work was later simplified by B. Kaufman, who evaluated the partition function using spinor representations. See also the article by Schultz, Mattis, and Lieb.

EXERCISES

1. Verify that the operators given in (13.4) and (13.5) are real forms of the algebras D_l and B_l respectively.

2. Show that $a_1^* \ldots a_l^* |0\rangle$ has the weight described in the text.

3. Verify that the algebra given in (13.6) is isomorphic to the real form (10.6) of $sp(2l, C)$.

4. Obtain boson realizations of the fundamental representations of the algebras A_l and C_l.

APPLICATIONS

CHAPTER 14

Applications

45. Completely Integrable Systems

The equations

$$u_t + u_{xxx} + uu_x = 0 \qquad \text{Korteweg–deVries}$$

$$u_{xt} = \sin u \qquad \text{Sine–Gordon}$$

$$iu_t + u_{xx} + 2|u|^2 u = 0 \qquad \text{Nonlinear Schrodinger}$$

have several remarkable features in common, among them

(i) a Hamiltonian structure.
(ii) an infinite number of conservation laws; all of them Hamiltonians in involution.
(iii) an associated spectral problem, invariant under the flow.
(iv) solvability by inverse scattering method.

In addition to the above three examples there are entire hierarchies of such "completely integrable Hamiltonian systems" based on any semi-simple Lie algebra. In this section we shall explain these ideas and give some further examples.

Let us begin by discussing the Korteweg–deVries (KdV for short) equation. Historically it was the first equation found to have such a remarkable set of properties. This was done in a series of papers by Kruskal, Zabusky, Miura, Gardner and Greene. (For a survey of the subject, see the book by Dodd et al.) We follow a presentation due to Lax [1]. Consider the Schrodinger operator

$$L = D^2 + (\tfrac{1}{6})u, \qquad D = \frac{\partial}{\partial x}.$$

It was shown that if u evolves according to the KdV equation then the family of operators $L(t) = D^2 + \frac{1}{6}u(x,t)$ remains unitarily equivalent. This means that there is a one parameter family of unitary operators $U(t)$ on the Hilbert space $L_2(-\infty, \infty)$ such that

$$U^*(t)L(t)U(t) = L(0).$$

Differentiating this identity with respect to time we obtain

$$L_t = [B, L], \tag{14.1}$$

where B is the infinitesimal generator of the family $U(t)$; that is $U_t = BU$. (Compare (14.1) with (4.3).) Since $L_t = \frac{1}{6}u_t$ we get the equation

$$u_t = 6[B, L].$$

This has to be interpreted as an operator equation; thus the commutator on the right must be a multiplication operator. This requires that the commutator $[L, B]$ contain no differentiations. The simplest example for a choice of B is the operator $B = D$. In that case $[B, L] = \frac{1}{6}u_x$; and the Lax equation (14.1) is simply $u_t = u_x$. This equation is the infinitesimal generator of translations, a one parameter group of transformations that clearly leaves the operator L invariant. The next case is far less obvious. We try to find a third order skew adjoint operator B (B must be skew adjoint in order to generate a group of unitary transformations) such that $[B, L]$ is a multiplication operator. Trying

$$B_1 = D^3 + bD + Db,$$

we find, after some computations, that for $b = \frac{1}{8}u$ the commutator $[B_1, L]$ becomes $(\frac{1}{24})(u_{xxx} + uu_x)$; and the KdV equation now follows from the Lax equations (14.1) if we take $B = -4B_1$.

The story does not stop here. In fact, there is an entire hierarchy of operators B, all of odd order, for which the commutator $[B, L]$ is a multiplication operator. Thus there is an infinite family of nonlinear evolution equations which leave the Schrodinger operator L invariant. This infinite family is closely tied to the infinite number of conservation laws for the KdV equation. We shall explain this now.

The KdV equation has a Hamiltonian structure, provided one looks at it in the right way. The Hamiltonian equations in the finite dimensional case in canonical coordinates are

$$\frac{dq_i}{dt} = \frac{\partial H}{\partial p_i} \qquad \frac{dp_i}{dt} = -\frac{\partial H}{\partial q_i}.$$

Combining the coordinates p_i and q_i into one variable $x = (q_1, \ldots p_1 \ldots)$ we may write the equations in the form

$$\frac{dx}{dt} = J\frac{\delta H}{\delta u} \qquad J = \begin{pmatrix} 0 & 1 \\ -1 & 0 \end{pmatrix} \tag{14.2}$$

where J is a $2n \times 2n$ matrix, $\mathbb{1}$ being the $n \times n$ identity matrix and 0 being the $n \times n$ zero matrix.

We can formally write the KdV equation in the same way, even though it is a partial differential equation. Consider functionals of the form

$$\mathfrak{H}(u) = \int H(u, u_x, \ldots) \, dx,$$

where H is a smooth function of u and its derivatives up to some finite order. Suppose $u(x, t)$ is a one parameter family of smooth functions in L_2 which vanishes along with all its derivatives as $|x| \to \infty$ and let us differentiate $\mathfrak{H}(u)$ with respect to t. We get

$$\frac{d\mathfrak{H}(u)}{dt} = \int \frac{\partial H}{\partial u} u_t + \frac{\partial H}{\partial u_x} u_{xt} + \cdots dx = \int \frac{\delta H}{\delta u} u_t \, dx$$

after integrating by parts. The expression $\delta H / \delta u$ is the Euler–Lagrange derivative of the density H, familiar in the calculus of variations:

$$\frac{\delta H}{\delta u} = \frac{\partial H}{\partial u} - \frac{d}{dx} \frac{\partial H}{\partial u_x} + \frac{d^2}{dx^2} \frac{\partial^2 H}{\partial u_{xx}} - \cdots .$$

The Euler–Lagrange derivative is the infinite dimensional analogue of the gradient.

We next find an infinite dimensional analogue of the matrix J in (14.2). Note that J is skew symmetric with respect to the Euclidean inner product on R^{2n}. Likewise, the operator D is skew symmetric with respect to the Hilbert space inner product

$$(u, v) = \int u(x) v(x) \, dx;$$

it plays the role of the operator J. The KdV equation can be written in the form

$$u_t = D(\delta H / \delta u) \quad \text{where} \quad H = \tfrac{1}{2} u_x^2 - \tfrac{1}{6} u^3.$$

Just as in the case of classical Hamiltonian mechanics there are Poisson brackets for the infinite dimensional case. They are given by

$$\{\mathfrak{H}, \mathfrak{B}\} = \int (\delta \mathfrak{H}/\delta u) D (\delta \mathfrak{B}/\delta u) \, dx,$$

and are called the *Gardner–Poisson* brackets. One must interpret $\{\mathfrak{H}, \mathfrak{B}\}$ as a functional on the infinite dimensional space L_2; just as in the finite dimensional case the Poisson bracket of two functions on the phase space is another function on the phase space.

It is clear that the Gardner–Poisson brackets are skew symmetric; this follows immediately from the skew-symmetry of D. What is less obvious is that they satisfy the Jacobi identity:

$$\{\{\mathfrak{F}, \mathfrak{Q}\}, \mathfrak{H}\} + \{\{\mathfrak{Q}, \mathfrak{H}\}, \mathfrak{F}\} + \{\{\mathfrak{H}, \mathfrak{F}\}, \mathfrak{Q}\} = 0. \tag{14.3}$$

(For a proof, see Olver or Sattinger [4].)

It is a simple matter to verify that, just as in the finite dimensional case, if $\mathfrak{F}$ is a functional and u evolves according to a Hamiltonian flow with Hamiltonian $\mathfrak{H}$, then

$$\frac{d\mathfrak{F}}{dt} = \frac{d}{dt} \int F(u)\,dx = \int (\delta F/\delta u) u_t\,dx$$

$$= \int (\delta F/\delta u) D(\delta H/\delta u)\,dx = \{\mathfrak{F}, \mathfrak{H}\},$$

just as in (4.1). In particular, any functional which is in involution with $\mathfrak{H}$ will be conserved under the flow. It turns out that there are an infinite number of such functionals which are in involution with the Hamiltonian which gives rise to the KdV equation. These are all related by the Lenard recursion formula:

$$DF_{n+1} = HF_n \quad \text{where} \quad H = (D^3 + \tfrac{2}{3}uD + \tfrac{1}{3}u_x). \tag{14.4}$$

It is not immediately clear that the F_n will be local, that is, that they will be functions of u and its derivatives; for at each stage we must "invert" the operator D on the formal vector space of algebraic expressions in u and its x-derivatives. For example, $D^{-1}(u)$ is not a local operator. We could write, of course,

$$D^{-1}(u) = \int_{-\infty}^{x} u(\xi)\,d\xi$$

but this operator is not local in u; that is, the support of $D^{-1}(u)$ is not contained in that of u. Nevertheless.

Theorem 14.1. *The recursion relation* (14.4) *is solvable to all orders. Each function* F_n *is a polynomial in* u *and its derivatives up to order* $n - 1$. *Moreover, each* F_n *is the Euler–Lagrange derivative of a functional* $\mathfrak{H}_n$. *The Hamiltonians* $\mathfrak{H}_n$ *are all in involution:* $\{\mathfrak{H}_n, \mathfrak{H}_m\} = 0$.

For a proof of this theorem, see Lax [2]. As a result of this theorem, each corresponding flow has an infinite number of conservation laws. The operator H gives rise to a second set of Poisson brackets

$$\{\mathfrak{F}, \mathfrak{B}\}_2 = \int (H\delta F/\delta u)(\delta G/\delta u)\,dx.$$

This second set of Poisson brackets is also skew symmetric and satisfies the Jacobi identity.

This remarkable set of properties of the KdV equation was at first thought to be very special, and therefore a singularity. But in fact it was soon discovered that there are many such non-linear evolution equations which have similar properties. In 1972 Zakharov and Shabat showed that the nonlinear Schrodinger equation is also completely integrable and has an associated eigenvalue problem which remains invariant under the flow. The isospectral problem in this case is closely related to the semi-simple Lie algebra $sl(2, C)$.

Consider the pair of operators

$$D_x(z) = \partial/\partial x - 2iz\sigma_3 - Q, \qquad Q = u\sigma_+ - u^*\sigma_-,$$

$$D_t = \partial/\partial t - F(z, Q)$$

where $F = 2A\sigma_3 + B\sigma_+ + C\sigma_-$. Here $\sigma_\pm$ and σ_3 are the matrices

$$\sigma_+ = \begin{pmatrix} 0 & 1 \\ 0 & 0 \end{pmatrix} \qquad \sigma_- = \begin{pmatrix} 0 & 0 \\ 1 & 0 \end{pmatrix} \qquad \sigma_3 = \frac{1}{2}\begin{pmatrix} 1 & 0 \\ 0 & -1 \end{pmatrix}$$

which generate the Lie algebra $sl(2, R)$. The coefficients A, B, and C of F are taken to be polynomials in u, u^* and their derivatives up to some finite order. The nonlinear Schrodinger equation is obtained by taking

$$A = 2iz^2 - iuu^* \qquad B = -2zu + iu_x \qquad C = 2zu^* + iu_x^*$$

and setting $[D_x, D_t] = 0$. In fact,

$$[D_x, D_t] = \sigma_+(u_t - iu_{xx} - 2iu^{*2}u) - \sigma_-(u_t^* + iu_{xx}^* + 2iu^2u^*).$$

Subsequently, Ablowitz, Kaup, Newell, and Segur obtained other completely integrable systems of equations for the isopectral operator $D_x(z)$ with $Q = p\sigma_+ + q\sigma_-$. (See the book by Dodd et al. for an extensive discussion.)

We now describe an analog of the Lenard scheme that generates Hamiltonian hierarchies of nonlinear evolution equations for any matrix representation of a semi-simple Lie algebra. We assume the potential Q takes its values in a semi-simple Lie algebra $\mathfrak{g}$. Let $\mathfrak{A}$ be the space of smooth matrix valued functions on $(-\infty, \infty)$ which take their values in $\mathfrak{g}$; and define an inner product on $\mathfrak{A}$ (in general, not positive definite) by

$$(P, Q) = \int \mathrm{Tr}(P(x)Q(x))\, dx. \tag{14.5}$$

Suppose $\mathfrak{F}$ is a functional on $\mathfrak{A}$ with density F:

$$\mathfrak{F}(Q) = \int F(Q, Q_x, \ldots)\, dx.$$

We denote the gradient of $\mathfrak{F}$ with respect to the inner product $(\ ,\ )$ by ∇F; thus

$$(d/dt)\mathfrak{F}(Q(t)) = \int \mathrm{Tr}(\nabla F(Q)(Q_t)\, dx.$$

As in the case of the KdV hierarchy there are two Poisson brackets relative to this inner product, namely

$$\{\mathfrak{F}, \mathfrak{B}\}_1 = \int \mathrm{Tr}([H, \nabla F](\nabla G))\, dx,$$

and   (14.6)

$$\{\mathfrak{F}, \mathfrak{B}\}_2 = \int \mathrm{Tr}((\partial(\nabla F)/\partial x - [Q, \nabla F])(\nabla G))\, dx.$$

Here ∇F and ∇G are the gradients of the densities F and G relative to the inner product on $\mathfrak{A}$, and $H \in \mathfrak{h}$. The recursive scheme is

$$i[H, F_{n+1}] = \partial F_n/\partial x - [Q, F_{n+1}]$$

$$F_0 = iK,$$ (14.7)

where $H, K \in \mathfrak{h}$. As in the case of the Lenard scheme, the recursion scheme is not *a priori* solvable to all orders; for the operator i ad H is only invertible on off diagonal matrices.

Theorem 14.2. *The hierarchy* (14.7) *is solvable to all orders. Each F_n is a polynomial in Q and its derivatives up to order $n - 1$, and is the gradient of a functional $\mathfrak{H}_n$. These Hamiltonians are all in involution with respect to both Poisson brackets $\{\ ,\ \}_1$ and $\{\ ,\ \}_2$ of* (14.6):

$$\{\mathfrak{H}_n, \mathfrak{H}_m\}_1 = \{\mathfrak{H}_n, \mathfrak{H}_m\}_2 = 0.$$

The corresponding evolution equations are

$$\partial Q/\partial t = i[H, F_{n+1}] = \partial F_n/\partial x - [Q, F_n].$$

The resulting Hamiltonian flows are parametrized by the matrix $K \in \mathfrak{h}$ and thus comprise an r (r = rank $\mathfrak{g}$) parameter family of commuting Hamiltonian flows.

For example, let us consider the first few terms of the hierarchy for the Lie algebra $sl(2, R)$:

$$i[\sigma_3, F_{j+1}] = \partial F_j/\partial x - [Q, F_j] \qquad F_0 = i\sigma_3,$$

$$Q = p\sigma_+ + q\sigma_-.$$

We have $i[\sigma_3, F_1] = -[Q, i\sigma_3]$; hence we take $F_1 = Q$. This gives the first evolution equations of the hierarchy:

$$\partial Q/\partial t = i[\sigma_3, Q],$$

or,

$$\partial p/\partial t = ip \qquad \partial q/\partial t = -iq.$$

At the next iteration we have $i[\sigma_3, F_2] = \partial F_1/\partial x - [Q, F_1] = \partial Q/\partial x - [Q, Q]$. Taking $F_2 = 2A_2\sigma_3 + B_2\sigma_+ + C_2\sigma_-$, we find $iB_2 = \partial p/\partial x$ and $iC_2 = \partial q/\partial x$. The term A_2 will be determined at the next stage. The equations at the second level of the hierarchy are therefore

$$\partial p/\partial t = \partial p/\partial x \qquad \partial q/\partial t = \partial q/\partial x.$$

At the next stage of the hierarchy we have

$$i[\sigma_3, F_3] = \partial F_2/\partial x - [Q, F_2] = (\partial/\partial x - \text{Ad } Q)\begin{bmatrix} A_2 & -ip_x \\ iq_x & -A_2 \end{bmatrix}$$

$$= 2(\partial A_2/\partial x - i(pq)_x)\sigma_3 + (-ip_{xx} + 2pA_2)\sigma_+ + (iq_{xx} - 2qA_2)\sigma_-.$$

Now the diagonal terms of the right hand side must vanish in order that it be in the range of $\text{Ad } \sigma_3$. Therefore we must take $A_2 = i(pq)$. Note that it is $\partial A_2/\partial x$ that must be determined, and it is given as the exact derivative $i(pq)_x$. Thus we obtain a polynomial in p and q for A_2, and A_2 is a *local* expression of p and q; that is, A_2 at the point x depends only on the values of p, q, and their x derivatives at x. The assertion of Theorem 14.1 is that this will be the case at all orders of the recursion procedure. Taking $A_2 = i(pq)$ we find that the equations at the third level of the hierarchy are

$$ip_t = p_{xx} - 2p^2q \qquad iq_t = -q_{xx} + 2pq^2.$$

The nonlinear Schrodinger equation is obtained by taking $p = -q^*$.

We now compute the first two equations in the hierarchy (14.7) for the Lie algebra $sl(3, R)$. We use the representation of $sl(3, R)$ given in §33. We assume the potential is Hermitian: $Q^* = Q$; thus

$$Q = \begin{pmatrix} 0 & Q_1 & Q_3 \\ Q_1^* & 0 & Q_2 \\ Q_3^* & Q_2^* & 0 \end{pmatrix}.$$

We shall write $Q = \sum_\alpha Q^\alpha E_\alpha$, where the sum runs over the non-zero roots α, and E_α are the root vectors. The condition $Q^* = Q$ is equivalent to $Q_{-\alpha} = Q_\alpha^*$. Writing $H = \sum_j h_j E_{jj}$, recall that the functionals ω_j were given by $\omega_j(H) = h_j$. The roots are $\omega_{12} = \omega_1 - \omega_2$, etc. We denote these by $\alpha = \omega_{12}$, $\beta = \omega_{23}$, $\gamma = \omega_{13}$. Therefore

$$[K, Q] = \sum_\alpha \alpha(K)Q^\alpha E_\alpha.$$

Taking $F_0 = iK$ we find that

$$F_1 = \sum_\alpha (\alpha(K)/\alpha(H))Q^\alpha E_\alpha;$$

and the first equations of the hierarchy are

$$\partial Q/\partial t = i[H, F_1] = i[K, Q],$$

hence

$$\partial Q_\alpha/\partial t = i\alpha(K)Q_\alpha.$$

The computations at the next stage are a little more involved. Let us set

$$\lambda_1 = \alpha(K)/\alpha(H) \qquad \lambda_2 = \beta(K)/\beta(H) \qquad \lambda_3 = \gamma(K)/\gamma(H).$$

Then

$$F_1 = \begin{pmatrix} 0 & \lambda_1 Q_1 & \lambda_3 Q_3 \\ \lambda_1 Q_1^* & 0 & \lambda_2 Q_2 \\ \lambda_3 Q_3^* & \lambda_2 Q_2^* & 0 \end{pmatrix}$$

and

$$[Q, F_1] = \begin{pmatrix} 0 & (\lambda_3 - \lambda_2)Q_2^* Q_3 & (\lambda_2 - \lambda_1)Q_1 Q_2 \\ (\lambda_3 - \lambda_2)Q_2 Q_3^* & 0 & (\lambda_1 - \lambda_3)Q_1^* Q_3 \\ (\lambda_2 - \lambda_1)Q_1^* Q_2^* & (\lambda_1 - \lambda_3)Q_1 Q_3^* & 0 \end{pmatrix}.$$

The equations at this stage of the hierarchy are $\partial Q/\partial t = i[H, F_2] = \partial F_1/\partial x - [Q, F_1]$, hence

$$\partial Q_1/\partial t = \lambda_1 \partial Q_1/\partial x + (\lambda_3 - \lambda_2)Q_2^* Q_3$$

$$\partial Q_2/\partial t = \lambda_2 \partial Q_2/\partial x + (\lambda_1 - \lambda_3)Q_1^* Q_3$$

$$\partial Q_3/\partial t = \lambda_3 \partial Q_3/\partial x + (\lambda_2 - \lambda_1)Q_1 Q_2,$$

and so forth. These are known as the three wave interaction equations. The potentials Q_i interact according to the root diagram for A_2 below:

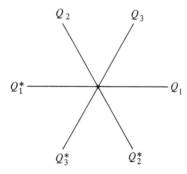

The computation of the next stage of the iteration is outlined in exercise 3 at the end of this section. The iteration at the next stage depends on the identity $\gamma = \alpha + \beta$ among the roots of the algebra A_2; but we get a system of non-linear partial differential equations, that is, a local equation. The assertion of Theorem 14.2 is that the same result is obtained for any semi-simple Lie algebra. That is, in carrying out the recursion relation (14.7) the terms are always local, so that the non-linear evolution equations are always partial differential equations, and not integro-differential equations.

The three wave interaction equations model the interaction of waves in nonlinear, conservative, dispersive media, and were first observed to be a completely integrable system by Zakharov and Manakov.

Kostant–Kirillov Symplectic Structures

Lie algebras play a fundamental role in the subject of completely integrable systems, whose surface we have barely touched upon in this introduction. They enter the subject naturally via the Kostant–Kirillov symplectic structure whose role was introduced into the subject independently by Adler, Kostant, and Symes. Consider the Hamiltonian system with Hamiltonian

$$H(x, y) = \frac{1}{2}\sum_j y_j^2 + \sum_j \exp\{x_j - x_{j+1}\}, \quad \text{with} \quad x_0 = x_{n+1} = 0.$$

H. Flaschka observed that under the transformation

$$a_i = \tfrac{1}{2}\exp\{(x_i - x_{i+1})/2\}, \quad i = 1, \ldots n - 1$$

$$b_i = \tfrac{1}{2}y_i$$

these equations could be written in the form of the Lax equations $dL/dt = [P, L]$, with

$$L = \begin{pmatrix} b_1 & a_1 & 0 & \cdot & \cdot & \\ a_1 & b_2 & a_2 & 0 & \cdot & \cdot \\ 0 & a_2 & b_3 & a_3 & 0 & \cdot \\ 0 & 0 & a_3 & b_4 & a_4 & 0 \\ 0 & 0 & \cdot & \cdot & \cdot & \end{pmatrix}$$

$$P = \begin{pmatrix} 0 & a_1 & 0 & \cdot & & \\ -a_1 & 0 & a_2 & 0 & \cdot & \\ 0 & -a_2 & 0 & a_3 & 0 & \cdot \\ 0 & 0 & -a_3 & 0 & a_4 & \end{pmatrix}.$$

The standard symplectic structure, in the $x - y$ variables, converted to the $a - b$ variables, leads to the relations $\{\operatorname{tr} L^j, \operatorname{tr} L^k\} = 0$; in other words, the Hamiltonians are $\operatorname{tr} L^j$, and these are in involution.

Kostant and Symes studied generalizations of this system of equations which were related to decompositions of semi-simple algebras. Kostant, for example, found completely integrable systems with Hamiltonians

$$H = \sum_j p_j^2/2m_j + U$$

where $U = r_1 \exp\{\varphi_1\} + r_2 \exp\{\varphi_2\} + \cdots r_n \exp\{\varphi_n\}$; the φ_j are linear functions of the q_j's:

$$\varphi_i = \sum_j a_{ij}q_j.$$

For special choices of the r_j and a_{ij}, having to do with the Dynkin diagram of a semi-simple Lie algebra, completely integrable finite dimensional Hamiltonian systems are obtained. For example, the Hamiltonian

$$H = (\tfrac{1}{2})\sum_j p_j^2 + \exp\{q_1 - q_2\} + \cdots + \exp\{q_{n-1} - q_n\} + \exp\{q_n\}$$

is related to the Lie algebra B_n, while

$$H = (\tfrac{1}{2})\sum_j p_j^2 + \exp\{q_1 + q_2\} + \exp\{q_2 - q_3\} + \exp\{q_3\}$$

$$+ \exp\{(q_4 - q_1 - q_2 - q_3)/2\}$$

is related to the exceptional Lie algebra F_4. Kostant obtains completely integrable systems for all the semi-simple Lie algebras and also obtains explicit solutions for these systems using representation theory. An independent analysis was also carried out by Symes.

Both used a symplectic structure which had been employed by Kirillov, together with certain direct sum decompositions of a Lie algebra into orthogonal subalgebras. Let $\mathfrak{g}$ be a Lie algebra and $\mathfrak{g}^*$ its dual relative to a nondegenerate bilinear form $\langle\ ,\ \rangle$ on $\mathfrak{g} \times \mathfrak{g}^*$. Let $\mathfrak{G}$ be a connected Lie group generated by $\mathfrak{g}$. The adjoint action of $\mathfrak{G}$ and its algebra $\mathfrak{g}$ on $\mathfrak{g}$ are given respectively by

$$\mathrm{Ad}\, g(y) = gyg^{-1}, \qquad \mathrm{ad}\, x(y) = [x, y], \qquad x, y \in \mathfrak{g}, \qquad g \in \mathfrak{G}.$$

By the duality $\langle\ ,\ \rangle$ these actions induce corresponding actions, called the co-adjoint actions and denoted by $\mathrm{Ad}'\, g$ and $\mathrm{ad}'\, x$, on the dual space $\mathfrak{g}^*$. These are defined by

$$\langle \mathrm{Ad}\, g(y), \alpha \rangle = \langle y, \mathrm{Ad}'\, g(\alpha) \rangle,$$

$$\langle \mathrm{ad}\, x(y), \alpha \rangle + \langle y, \mathrm{ad}'\, x(\alpha) \rangle = 0.$$

where $\alpha \in \mathfrak{g}^*$, $x, y \in \mathfrak{g}$, and $g \in \mathfrak{G}$. In the case of a semi-simple Lie algebra we may take $\langle\ ,\ \rangle$ to be the Killing form; in that case $\mathfrak{g}$ and $\mathfrak{g}^*$ are identified and the adjoint and co-adjoint action agree.

The co-adjoint orbit through $\alpha \in \mathfrak{g}^*$ is $M_\alpha = \{\mathrm{Ad}'\, g(\alpha)|g \in \mathfrak{G}\}$. The tangent space to M_α at α is given by $TM_\alpha = \{\mathrm{ad}'\, x(\alpha)|x \in \mathfrak{g}\}$. The co-adjoint orbits are symplectic manifolds with symplectic form ω_α at α given by

$$\omega_\alpha(\mathrm{ad}'\, x_1(\alpha), \mathrm{ad}'\, x_2(\alpha)) = \langle [x_1, x_2], \alpha \rangle.$$

This is the Kirillov symplectic structure: ω_α is a non-degenerate 2-form at all points of the co-adjoint orbit. The gradient of a smooth function $F: \mathfrak{g}^* \to R$ relative to the bilinear form $\langle\ ,\ \rangle$ is defined by

$$(d/dt)F(\alpha(t)) = \langle \nabla F(\alpha), d\alpha/dt \rangle$$

where $\alpha(t)$ is a curve in $\mathfrak{g}^*$, and $\nabla F: \mathfrak{g}^* \to \mathfrak{g}$. Equivalently we may say that ∇F is defined by $dF(Y) = \langle \nabla F, Y \rangle$, where Y is a tangent vector on $\mathfrak{g}^*$. The *Kirillov Poisson bracket* is

$$\{F, G\}(\alpha) = \langle [\nabla F(\alpha), \nabla G(\alpha)], \alpha \rangle.$$

It was proved by Kirillov that this expression is in fact a Poisson bracket, that is, that it is skew symmetric and satisfies the Jacobi identity.

Given a function H on $\mathfrak{g}^*$ the Kirillov form ω induces a Hamiltonian vector field X_H on $\mathfrak{g}^*$ defined implicitly by $\omega(X_H, Y) = YH$, where $Y \in \mathfrak{g}^*$. We compute X_H as follows. Since X_H and Y are tangent vectors to the coadjoint orbit they may be written as $Y = \mathrm{ad}' x_2(\alpha)$ and $X_H = \mathrm{ad}' x_1(\alpha)$. Then

$$YH = \langle \nabla H, Y \rangle = \langle \nabla H, \mathrm{ad}' x_2(\alpha) \rangle = -\langle [x_2, \nabla H], \alpha \rangle$$
$$= \omega(X_H, Y) = \langle [x_1, x_2], \alpha \rangle.$$

Comparing these two expressions we find $x_1 = \nabla H$, since x_2 is arbitrary. Hence the Hamiltonian vector field is given by

$$X_H(\alpha) = \mathrm{ad}' \nabla H(\alpha).$$

Kostant and Symes use the Kirillov bracket in the following way. Suppose that the real Lie algebra $\mathfrak{l}$ is the direct sum of $\mathfrak{g}$ and $\mathfrak{h}$, that is $\mathfrak{l} = \mathfrak{g} + \mathfrak{h}$, where $\mathfrak{g}$ and $\mathfrak{h}$ are both subalgebras. By the duality, $\mathfrak{l}^* = \mathfrak{g}^* + \mathfrak{h}^*$, where $\mathfrak{g}^*$ is identified with $\mathfrak{h}$, and $\mathfrak{h}^*$ is identified with $\mathfrak{g}$. Thus $\mathfrak{g}$ naturally inherits the Kostant–Kirillov symplectic structure given by

$$\{F, G\}(\alpha) = \langle [\Pi_\mathfrak{g} \nabla F(\alpha), \Pi_\mathfrak{g} \nabla G(\alpha)], \alpha \rangle, \qquad \alpha \in \mathfrak{g},$$

where $\Pi_\mathfrak{g}$ is the projection on $\mathfrak{g}$ along $\mathfrak{h}$. Let $\mathfrak{S} \subset C^\infty(\mathfrak{l}^*)$ be the class of Ad' invariant functions which are constant on the coadjoint orbits of $\mathfrak{L}$ (the connected Lie group generated by $\mathfrak{l}$) on $\mathfrak{l}^*$.

Theorem 14.3 (Kostant, Symes). *The functions of the family $\mathfrak{S}$ of Ad'-invariant functions, restricted to the subspace $\mathfrak{g}^*$, are in involution with respect to the bracket given above: $\{F_1, F_2\} = 0$ for $F_1, F_2 \in \mathfrak{S}|_{\mathfrak{g}^*}$. The Hamiltonian vector field induced by $F \in \mathfrak{S}|_{\mathfrak{g}^*}$ is given by*

$$H_F(\alpha) = \mathrm{ad}' \Pi_\mathfrak{g} \nabla F(\alpha)(\alpha) = -\mathrm{ad}' \Pi_\mathfrak{h} \nabla F(\alpha)(\alpha).$$

When $\mathfrak{l}$ is semi-simple we may use the Killing form. In that case $\mathfrak{l}$ and its dual are identified, and $\mathrm{ad} = \mathrm{ad}'$. In this case the Poisson bracket is given by

$$\{F_1, F_2\}(\alpha) = K(\alpha, [\Pi_\mathfrak{g} \nabla F_1, \Pi_\mathfrak{g} \nabla F_2]),$$

and the Hamiltonian vector field is now

$$H_F(\alpha) = [\Pi_\mathfrak{g} \nabla F(\alpha), \alpha] = [\alpha, \Pi_\mathfrak{h} \nabla F(\alpha)].$$

The Lax equations are then $dx/dt = [\alpha, \Pi_\mathfrak{h} \nabla F(\alpha)]$.

The equations of the Toda lattice are obtained by taking $\mathfrak{l} = sl(n, R)$, $\mathfrak{g}$ the subalgebra of lower triangular matrices with non-zero diagonal entries and $\mathfrak{h}$ the dual of $\mathfrak{g}$ relative to the bilinear form $\langle E, F \rangle = \mathrm{Tr}\, EF$. Then $\mathfrak{h}$ is the algebra of strictly upper triangular matrices. (See Adler.)

Adler extended the Kirillov–Kostant–Symes approach to infinite dimensional Lie algebras and treated the KdV hierarchy (and generalizations) by applying this formalism to an algebra of pseudo-differential operators connected with the formal calculus of variations.

EXERCISES

1. Compute the expression for the gradient of a functional relative to the inner product (14.6). Let E_α be a basis of $\mathfrak{g}$; and let $K_{\alpha\beta} = \text{Tr}\, E_\alpha E_\beta$ be the components of the metric tensor. We write $Q = Q^\alpha E_\alpha$. Show that

$$d\mathfrak{F}(Q)/dt = \int (\delta F/\delta Q^\alpha)(\partial Q^\alpha/\partial t)\, dx,$$

hence that $(\nabla F)^\alpha = K^{\alpha\beta}(\delta F/\delta Q^\beta)$, where $K^{\alpha\beta}$ is the inverse of the matrix $K_{\alpha\beta}$.

2. Let $\Psi(x, t)$ satisfy the differential equation $D_x \Psi = 0$, and put $\Psi = m(x, t)\exp\{ixzH\}$. Show that m satisfies the equation

$$\partial m/\partial x = iz[H, m] + Qm.$$

For any fixed matrix K, show that $F = mKm^{-1}$ satisfies the equation

$$\partial F/\partial x = iz[H, F] + [Q, F].$$

If F has the expansion $F = \sum_n z^{-n} F_n$, show that the F_n satisfy the recursion scheme (14.7).

3. Carry out one more iteration step in the $sl(3, R)$ hierarchy and find two Hamiltonians for the three wave interaction, one for each of the two Hamiltonian structures i Ad H and $\partial/\partial x - \text{Ad}\, Q$. Show that one Hamiltonian is $H_1 = \sum_j \lambda_j Q_j Q_j^*$ and that the second Hamiltonian is

$$H_2 = (\tfrac{1}{2})\sum_j \lambda_j(Q_j^* \partial Q_j/\partial x - Q_j \partial Q_j^* \partial x) + \Omega(Q_1^* Q_2^* Q_3 + Q_1 Q_2 Q_3^*)$$

where

$$\Omega = \frac{\lambda_1 - \lambda_2}{h_1 - h_3} = \frac{\lambda_2 - \lambda_3}{h_1 - h_2} = \frac{\lambda_3 - \lambda_1}{h_2 - h_3}.$$

The above identity in the three expressions for Ω is a consequence of the relation $\alpha + \beta = \gamma$ among the roots of the Lie algebra A_2. This identity is needed in the iteration step in solving for F_2.

46. Spontaneous Symmetry Breaking

The term "spontaneous symmetry breaking" originally arose in elementary particle physics to describe the fact that the vacuum state had less symmetry than the group invariance of the equations; but the notion also has relevance for problems in classical mechanics, specifically in the realm of bifurcation theory. When physical systems become unstable, the new solutions which appear often possess a lower isotropy symmetry group.

For example, a layer of fluid heated from below becomes unstable when the temperature drop across the layer exceeds a certain critical value; and the flow patterns which arise often display a cellular (i.e. crystallographic) structure.

In carefully controlled experiments, a homogeneous layer of fluid in a circular dish develops hexagonal flow patterns. Similar hexagonal patterns are often seen in nature on the surface of cooled rocks, and especially in basalt formations. Such crystallographic patterns furnish excellent examples of "spontaneously broken symmetry." Prior to the onset of instability the solutions are invariant under some group of transformations; but after instability sets in, the new solutions are invariant only under some subgroup. The governing equations themselves continue to be invariant under the full transformation group; thus the symmetry breaking is spontaneous.

Such examples of spontaneously broken symmetry are pervasive in nature, and occur in the context of phase transitions, buckling of plates, convection in fluids, as well as in biological systems. (See Golubitsky and Schaeffer, Marsden, or Sattinger.)

Let us discuss a simple mathematical example to illustrate the ideas. Consider the partial differential equation

$$G(\lambda, u) = \Delta u + \lambda u + f(u) = 0 \tag{14.8}$$

on the unit sphere in n dimensions, where Δ is the usual Laplacian on the sphere. Let O be a matrix in $SO(n)$, and let $\Gamma_O u(x) = u(O^{-1}x)$. Then Γ_O is a representation of $SO(n)$ on functions defined on the unit sphere, and equation (14.8) is *equivariant* with respect to Γ_O; that is, $\Gamma_O G(\lambda, u) = G(\lambda, \Gamma_O u)$. We assume $f(0) = 0$, so that $u = 0$ is always a solution of (14.8). We seek other solutions near 0 which *bifurcate* from this trivial solution. It is well-known that such bifurcation may take place near the eigenvalues of Δ, that is, for $\lambda = \lambda_j$ where

$$\Delta\varphi_j + \lambda_j\varphi_j = 0.$$

Due to the rotational invariance of the Laplacian, the eigenvalues in general have multiplicity greater than 1. For example, in R^3 the eigenvalue λ_j has multiplicity $2j + 1$. The original equation is, of course, an infinite dimensional problem; it is posed on some space of functions defined on the unit sphere. A standard method in bifurcation theory, known as the *Lyapounov–Schmidt* method, allows one in many cases to reduce infinite dimensional problems on a Banach space to finite dimensional ones.

We state the principal results for the general case here. Let $G(\lambda, u)$ be an analytic mapping from a complex Banach space $\mathfrak{E}$ to $\mathfrak{F}$. This means that G has a Frechet derivative with respect to u everywhere on $\mathfrak{E}$. For simplicity assume that $G(\lambda, 0) = 0$. Assume $G_u(\lambda, u)$, the Frechet derivative of G with respect to u, is a Fredholm operator with finite dimensional kernel $\mathfrak{N}$ and closed range $\mathfrak{R}$, with codim $\mathfrak{R} = \dim \mathfrak{N}$. We further assume that G is equivariant with respect to a representation Γ of a compact group $\mathfrak{G}$ acting on $\mathfrak{E}$ and $\mathfrak{F}$; thus $\Gamma G(\lambda, u) = G(\lambda, \Gamma u)$. Under these hypotheses:

Theorem 14.4. *Suppose $G_u(\lambda, 0)$ has a nontrivial kernel $\mathfrak{N}$ for $\lambda = \lambda_0$. Then $\mathfrak{N}$ is invariant under Γ; and Γ restricted to $\mathfrak{N}$ is a finite dimensional representation* of $\mathfrak{G}$. *All solutions of* $G(\lambda, u) = 0$ *in a neighborhood of* $(\lambda_0, 0)$ *can be obtained by solving a system of finite dimensional equations*

$$F(\lambda, \chi) = 0, \qquad F: \mathfrak{N} \to \mathfrak{N} \tag{14.9}$$

where $\chi \in \mathfrak{N}$. *The branching equations* (14.9) *are covariant with respect to the representation* Γ *restricted to the kernel* $\mathfrak{N}$. *Denoting this representation by* Γ_0 *we have* $\Gamma_0 F(\lambda, \chi) = F(\lambda, \Gamma_0 \chi)$.

The problem of finding nontrivial solutions of equation (14.8) is thus reduced to solving finite systems of algebraic equations in the vicinity of the branch point. These algebraic equations are in turn covariant with respect to the representation Γ_0. In the case of the rotation group $SO(3)$, for example, the relevant representations are the finite dimensional irreducible representations D^l.

While the covariance of the bifurcation equations can be used to gain considerable information about their algebraic structure, it still is often inadequate to resolve completely the bifurcation problem. For example, in the case of the rotation group $SO(3)$ the bifurcation problem has been completely solved only for the first two eigenvalues. In the case of the higher eigenvalues some solutions are known, but a complete analysis of the bifurcating solutions is so far unknown.

We discuss some ideas of *L. Michel* concerning the existence of bifurcating solutions under fairly general circumstances. Let Γ act on the vector space V; and let $F(\lambda, v) = \lambda v - f(\lambda, v)$ be a set of bifurcation equations, covariant with respect to a finite dimensional representation Γ of a group $\mathfrak{G}$: $\Gamma F(\lambda, v) = F(\lambda, \Gamma v)$. We assume $f(\lambda, 0) = 0$ and that f consists only of terms which vanish to higher than second order.

Definition 14.5. Let $v_0 \in V$. We say that v_0 is a direction of symmetry breaking if every mapping $F(\lambda, v)$ has a nontrivial solution along the direction v_0. In other words, for every covariant mapping $F = \lambda v - f(\lambda, v)$ satisfying the above conditions there exists a solution of the form $v(\lambda) = \tau(\lambda) v_0$, where $\tau(\lambda)$ is some continuous function vanishing at 0.

Given a vector $v_0 \in V$, its isotropy subgroup is defined to be the subgroup $\mathfrak{G}_0 = \{g \in \mathfrak{G} | \Gamma_g v_0 = v_0\}$. If Γ is a representation of $\mathfrak{G}$ on V, then Γ restricted to the subgroup $\mathfrak{G}_0$ decomposes into a direct sum of irreducible representations of $\mathfrak{G}_0$. We write

$$\Gamma|_{\mathfrak{G}_0} = \sum_v a_v \Gamma^v$$

where a_v are non-negative integers and Γ^v are the irreducible representations of $\mathfrak{G}_0$. There is a corresponding decomposition of the vector space V into a direct sum of subspaces V^v, each irreducible under Γ^v. We denote the identity representation by Γ^0; thus $\Gamma_g^0 v_0 = v_0$ for each $v_0 \in V^0$ and $g \in \mathfrak{G}_0$.

The following theorem has been proved in various forms by Cicogna, Michel, and Vander bauwhede.

Theorem 14.6. *Let Γ be a finite dimensional representation of a group $\mathfrak{G}$; let v_0 have isotropy subgroup $\mathfrak{G}_0$; and let a_0 be the number of times $\Gamma|_{\mathfrak{G}_0}$ contains the identity representation. If $a_0 = 1$ then v_0 is a direction of symmetry breaking.*

PROOF. Since f is covariant we have $\Gamma_g f(\lambda, r v_0) = f(\lambda, \Gamma_g r v_0) = f(\lambda, r v_0)$ for all $g \in \mathfrak{G}_0$. Since $\Gamma|_{\mathfrak{G}_0}$ contains the identity representation only once, $f(\lambda, r v_0)$ must be a scalar multiple of v_0 for all values of λ and r. Thus we must have $f(\lambda, r v_0) = \sigma(\lambda, r) v_0$ for all λ and r, σ being a smooth function; and the bifurcation equation reduces to a one-dimensional problem $\sigma(\lambda, r) = 0$. From the properties of f, $\sigma(\lambda, r) = r\sigma_1(\lambda, r)$, where $\partial \sigma_1 / \partial \lambda(0, 0) \neq 0$. The non-trivial solutions are obtained by solving $\sigma_1(\lambda, r) = 0$ by the implicit function theorem. $\qquad \square$

Let us show how to apply this theorem in practice. Suppose H is a subgroup of $SO(3)$. We want to count the number of times the representation D^l of $SO(3)$ contains the identity representation when restricted to the subgroup H. (If D^l acts on a vector space V and v_0 is fixed under Γ_h for all $h \in H$, then H is contained in the isotropy subgroup for v_0. We shall assume that H is precisely the isotropy subgroup of v_0.) We determine this number as follows. The representation D^l can be obtained from the symmetric tensor product $(D^{1/2})_s^{\otimes 2l}$ as follows. The representation $D^{1/2}$ is given by

$$D^{1/2}(u, v) = (au + bv, -b^*u + a^*v),$$

where

$$A = \begin{pmatrix} a & b \\ -b^* & a^* \end{pmatrix}$$

is an element of $SU(2)$. The symmetric tensor product $(D^{1/2})_s^{\otimes 2l}$ acts on the space of homogeneous polynomials spanned by $\{u^{2l}, u^{2l-1}v, \ldots v^{2l}\}$. For example, $u^{2l} \to (au + bv)^{2l}$, etc. We leave it to the reader to show that the resulting representation is irreducible. It is therefore a $2l + 1$ dimensional representation of the covering group $SU(2)$, and therefore a representation of $SO(3)$. Since l is an integer, it is a single-valued representation of $SO(3)$. In fact, if the matrix A is replaced by $-A$, the transformed polynomials are left the same. So it must be D^l.

The number of invariants of $D^l|_H$ is therefore the equal to the number of invariants of $(D^{1/2})_s^{\otimes 2l}|_{H'}$, where H' is the pre-image of H in $SU(2)$. The latter number may be calculated by the Molien function:

Lemma 14.7. *Let Γ be a finite dimensional representation of a compact group $\mathfrak{G}$; and let $(\Gamma^{\otimes n})_s$ be the n-fold symmetric tensor product. Then the number of invariants is given by the coefficient c_n of z^n in the generating function*

$$\sum_n c_n z^n = 1/|\mathfrak{G}| \int_{\mathfrak{G}} \det(1 - z\Gamma(g))^{-1} \, d\mu(g) \tag{14.10}$$

where $d\mu(g)$ is the invariant measure on the group.

The generating function constructed in Lemma 14.6 is called the *Molien function*; it was discussed briefly in Chapter 7.

PROOF. Let Γ act on the vector space V. Every representation of a compact group is equivalent to a unitary representation, so we assume Γ is a unitary representation of $\mathfrak{G}$. For fixed $g \in \mathfrak{G}$ let $v_1, \ldots v_k$ be a basis of eigenvectors of $\Gamma(g)$ with eigenvalues $\lambda_1, \ldots, \lambda_k$. The vector space $(V^{\otimes n})_s$ is spanned by the tensor products

$$|m_1, \ldots m_k\rangle = 1/n! \sum_{\pi \in S_n} v_{\pi(i_1)} \otimes v_{\pi(i_2)} \otimes \cdots \otimes v_{\pi(i_n)}$$

where the vector v_i occurs m_i times, and $m_1 + \cdots + m_k = n$. The sum on the right is a completely symmetric vector in $V^{\otimes n}$. The action of $\Gamma^{\otimes n}$ is

$$\Gamma^{\otimes n}|m_1, \ldots m_k\rangle = (\lambda_1^{m_1} \lambda_2^{m_2} \ldots \lambda_k^{m_k})|m_1, \ldots m_k\rangle.$$

Therefore the character of the representation $\Gamma^{\otimes n}$ is given by

$$\chi_{(n)}(g) = \mathrm{Tr}(\Gamma^{\otimes n})_s(g) = \sum_{m_1 + \cdots m_k = n} (\lambda_1^{m_1} \lambda_2^{m_2} \ldots \lambda_k^{m_k}).$$

Multiplying by z^n and summing over n we get

$$\sum_{n=0}^{\infty} z^n \chi_{(n)}(g) = \sum_{n=0}^{\infty} z^n \sum_{m_1 + \cdots m_k = n} (\lambda_1^{m_1} \lambda_2^{m_2} \ldots \lambda_k^{m_k})$$

$$= \sum_{m_1, \ldots m_k = 0}^{\infty} (z\lambda_1)^{m_1} (z\lambda_2)^{m_2} \ldots (z\lambda_k)^{m_k}$$

$$= \prod_{j=1}^{k} (1 - z\lambda_j)^{-1}$$

$$= \det(1 - z\Gamma(g))^{-1}.$$

It is a standard fact from representation theory (see Miller, for example) that the number of times a representation Γ contains the identity representation is given by

$$\int_{\mathfrak{G}} \chi(g) \, d\mu(g)$$

where $d\mu(g)$ is the normalized invariant measure and χ is the character of Γ: $\chi(g) = \mathrm{Tr}\,\Gamma(g)$. Therefore upon integration we obtain (14.10). $\square$

We now apply this technique to gain information about the breaking of rotational symmetry. We calculate the Molien function for the representation $D^{1/2}$ restricted to the subgroup O', where O' is the group that covers the octahedral group O. The group O' has 48 elements in 8 conjugacy classes (see Hamermesh, p. 363) as follows:

1	1	8	8	6	6	6	12
I	R	C_3	C_3^2	C_4	C_4^3	C_4^2	C_2
		$C_3^2 R$	$C_3 R$	$C_4^3 R$	$C_4 R$	$C_4^2 R$	$C_2 R$

Here I is the 2×2 identity matrix; $R = -I$; and C_2, C_3 and C_4 are two, three, and four-fold rotations, that is, elements of orders 2, 3, and 4. The notation C_4^2 denotes the square of a four-fold rotation. For example, a three-fold rotation about the z axis is a rotation through $2\pi/3$ and lifts to the 2×2 matrix

$$\begin{pmatrix} e^{i\pi/3} & 0 \\ 0 & e^{-i\pi/3} \end{pmatrix}.$$

The number of elements in each conjugacy class is given by the digit on the top line. The number is divided equally between the two types of elements in each class. For example, there are four elements of the form C_3 and four of the form $C_3^2 R$. The Molien function for this representation is

$$M_{1/2}(z) = (1/48) \sum_{g_j \in O'} \{(1 - z\lambda_j)(1 - z\lambda_j^*)\}^{-1}$$

where λ_j and λ_j^* are the eigenvalues of the element g_j. This sum reduces to a sum over the conjugacy classes, since elements in the same conjugacy class have the same eigenvalues. Let $\varepsilon = e^{i\pi/3}$ and $\alpha = e^{i\pi/4}$. For a three fold rotation the eigenvalues are $\pm\varepsilon$ while for an element of the form $C_3^2 R$ the eigenvalues are $\pm\varepsilon^2$.

The element I contributes the term $1/(1 - z)^2$; the contribution from R is $1/(1 + z)^2$; the contributions from C_3 and $C_3^2 R$ are both $\{(1 - z\varepsilon)(1 - z\varepsilon^*)\}^{-1}$; the contributions from C_3^2 and $C_3 R$ are $\{(1 + z\varepsilon)(1 + z\varepsilon^*)\}^{-1}$; and so forth. The Molien function is therefore

$$M_{1/2}(z; O') = (1/48)[(1 - z)^{-2} + (1 + z)^{-2} + 8\{(1 - z\varepsilon)(1 - z\varepsilon^*)\}^{-1}$$
$$+ 8\{(1 + z\varepsilon)(1 + z\varepsilon^*)\}^{-1} + 6\{(1 - z\alpha)(1 - z\alpha^*)\}^{-1}$$
$$+ 6\{(1 + z\alpha(1 + z\alpha^*)\}^{-1} + 18(1 + z^2)^{-1}].$$

After some algebraic manipulations this function can be simplified to

$$M_{1/2}(z; O')) = \frac{1 + z^{18}}{(1 - z^8)(1 - z^{12})} \tag{14.11}$$

$$= 1 + z^8 + z^{12} + z^{18} + z^{20} + 2z^{24} + z^{26} + z^{28}$$
$$+ z^{30} + 2z^{32} + z^{34} + 2z^{36} + \cdots.$$

We saw that the number of invariants of $D^l|_O$ (that is, the number of times $D^l|_O$ contains the identity representation) was equal to the number of invariants of $(D^{1/2})^{2l}|_{O'}$. Therefore, the number of invariants of $D^l|_O$ is equal to the coefficient of z^{2l} in the Molien function $M_{1/2}(z)$. For example, $D^l|_O$

contains the identity representation precisely once for $l = 4, 6, 9, 10, 13, 14,$ 15, 17, 19, and 23.

By similar methods one may obtain the Molien function for $D^{1/2}$ restricted to the other point groups of $SU(2)$. For the tetrahedral and icosahedral groups we get

$$M_{1/2}(z; T') = \frac{1 + z^{12}}{(1 - z^6)(1 - z^8)} \tag{14.12}$$

$$M_{1/2}(z; Y') = \frac{1 + z^{30}}{(1 - z^{12})(1 - z^{20})}. \tag{14.13}$$

A second important class of symmetry breaking bifurcations is the bifurcation of time periodic motions in systems invariant under a group of spatial transformations. Thus, we consider systems of differential equations of the form

$$\partial u/\partial t - H(\lambda, u) = 0. \tag{14.14}$$

Again, we assume that u takes its values in a Banach space $\mathfrak{E}$ and that the equations are equivariant with respect to a representation Γ of a compact group $\mathfrak{G}$ acting on $\mathfrak{E}$. The full symmetry group of (14.14) is therefore $\mathfrak{R} \times \mathfrak{G}$. As λ crosses a critical value λ_0, time periodic solutions of (14.14) can bifurcate from the zero solution. Such solutions will generally have a subgroup of $\mathbb{Z} \times \mathfrak{G}$ as a symmetry group, and will appear physically as waves. The determination of the frequency of the bifurcating solutions must be determined as part of the solution of the problem. We therefore set $s = \omega t$ and $u(t) = v(\omega t)$ in (14.14), obtaining the equation

$$\omega \partial v/\partial s - H(\lambda, v) = 0,$$

for the 2π periodic function $v(s)$. Under appropriate assumptions on the operator H, the Lyapounov–Schmidt method can again be applied, provided that the linear operator

$$J = \partial v/\partial s - L_0, \qquad L_0 = H_v(\lambda_0, 0)$$

satisfies the Fredholm alternative on an appropriate Banach space of 2π periodic functions.

The kernel of the operator J takes the form

$$\mathfrak{R} = \{e^{is}\Psi + e^{-is}\Psi^* | L_0\Psi = i\Psi, \qquad L_0^*\Psi^* = -i\Psi^*\}.$$

The bifurcation equations may be written

$$F(\lambda, \omega, \Psi, \Psi^*) = p(\lambda, \omega, \Psi, \Psi^*)e^{is} + (p(\lambda, \omega, \Psi, \Psi^*))^* e^{-is}$$

where $p(\lambda, \omega, \Psi, \Psi^*) \in \ker(iI - L_0)$, and $p^* \in \ker(iI + L_0)$. The action of the group $\mathfrak{G}$ on $\ker(iI \pm L_0)$ is given by Γ restricted to these subspaces. Since F is equivariant, we have $\Gamma(g)p(\lambda, \omega, \Psi, \Psi^*) = p(\lambda, \omega, \Gamma(g)\Psi, \Gamma(g)\Psi^*)$. Because of the special nature of the bifurcation equations, they reduce to half the number, viz.

$$p(\lambda, \omega, \Psi, \Psi^*) = 0.$$

By examining the Lyapounov–Schmidt procedure it can be determined that the leading terms of p are

$$p(\lambda, \omega, \Psi, \Psi^*) = i\omega - \gamma(\lambda) + q(\lambda, \omega, \Psi, \Psi^*)$$

where q is equivariant and consists of higher order terms, and $\gamma(\lambda)$ is the principal eigenvalue of $H_u(\lambda, 0)$: in other words, $H_u(\lambda, 0)\varphi(\lambda) = \gamma(\lambda)\varphi(\lambda)$.

The ideas discussed in this section provide only a small introduction to the subject of symmetry breaking in bifurcation problems. For further study see the references cited in this section.

EXERCISES

1. Obtain the form (14.11) for the Molien function $M_{1/2}(z; O')$.

2. Derive the Molien functions (14.12) and (14.13). The group T' has 24 elements in 7 conjugacy classes:

1	$\{I\}$	4	$\{C_3^2\}$
1	$\{R\}$	4	$\{C_3^2 R\}$
4	$\{C_3\}$	6	$\{C_2, C_2 R\}$
4	$\{C_3 R\}$		

For the icosahedral group Y the group Y' has 120 elements in 9 conjugacy classes, as follows:

1	$\{1\}$	12	$\{C_{4\pi/5}R, C_{6\pi/5}\}$
1	$\{R\}$	20	$\{C_{2\pi/3}, C_{4\pi/3}R\}$
12	$\{C_{2\pi/5}, C_{8\pi/5}R\}$	20	$\{C_{2\pi/3}R, C_{4\pi/3}\}$
12	$\{C_{2\pi/5}R, C_{8\pi/5}\}$	30	$\{C_\pi, C_\pi R\}$
12	$\{C_{4\pi/5}, C_{6\pi/5}R\}$		

Bibliography

Adler, M.

On a trace functional for formal pseudo-differential operators and the symplectic structure of the Korteweg–deVries type equations, *Inv. Math.* **50** (1979), 219–248.

Ado, I.D.

The representation of Lie algebras by matrices, *Uspekhi Mat. Nauk.* (N.S.) **2**, No. 6 (1947), 159–173. (Amer. Math. Soc. Transl. Series 1, **9**, p. 308, 1949.)

Arnold, V.I.

[1] Sur la geometrie differentielle des groupes de Lie de dimension infinie et ses applications a l'hydrodynamique des fluides parfaites, *Ann. Inst. Fourier*, Grenoble, **16** (1966), 319–361.

[2] *Mathematical Methods of Classical Mechanics*, Springer-Verlag, New York, 1978.

Biedenharn, L.C. and Van Dam.

Quantum Theory of Angular Momentum, Academic Press, New York, 1975.

Birkhoff, G.

Lie groups simply isomorphic with no linear group, *Bull. Amer. Math. Soc.* **42** (1936), 883–888.

Blumen, G.W. and Cole, J.D.

Similarity Methods for Differential Equations, Springer-Verlag Applied Mathematical Sciences, vol. 13, New York, 1974.

Brauer, R. and Weyl, H.

Spinors in n dimensions, *Amer. Jour. Math.* **57** (1935), 425–449.

Caratheodory, C.

Calculus of Variations and Partial Differential Equations of the First Order, Vol. 1, Holden-Day, San Francisco, 1965.

Cartan, E.

[1] Sur la structure des groupes de transformations finis et continus, These; Paris, Nony 1894 (Oeuvres Completes, Partie I. tome 1, pp. 137–287.

[2] La structure des groupes de transformations continus et la theorie du triedre mobile, Oevres Completes, P. III, t.1, Gauthiers-Villars, Paris, 1955.

[3] Theorie des groupes finis et continus et la geometrie differentielle, traites par la methode du repere mobile, Gauthier-Villars, Paris, 1937.

Chevalley, C.
Theory of Lie Groups I, Princeton University Press, 1946.

Cicogna, G.
Symmetry breakdown from bifurcations, *Lettere al Nuovo Cimento*, **31** (1981), 600–602.

Dirac, P.A.M.
[1] *The Principles of Quantum Mechanics*, Oxford University Press, 1935.
[2] Quantized singularities in the electromagnetic field, *Proc. Royal Soc. of London*, A, **133** (1931) 60–72.

Dodd, R.K., Eilbeck, J.C., Gibbon, J.D. and Morris, H.C.
Solitons, Academic Press, New York, 1982.

Dynkin, E.B.
The structure of semisimple Lie algebras, *Uspekhi Mat. Nauk.* (N.S.) **2** (1947) 59–127 (Amer. Math. Soc. Translations Series 1, **9**, Providence, 1962.)

Ebin, D.G. and Marsden, J.
Groups of diffeomorphisms and the motion of an incompressible fluid, *Annals of Mathematics* **92** (1970), 102–163.

Feynman, R.P., Leighton, R.B. and Sands, M.
The Feynman Lectures on Physics, Addison-Wesley, Reading, Massachusetts, 1975.

Flanders, H.
Differential Forms, Academic Press, New York, 1963.

Georgi, H.
Lie Algebras in Particle Physics, Benjamin, Reading, Massachusetts, 1982.

Gibson, W.M. and Pollard, B.R.
Symmetry Principles in Elementary Particle Physics, Cambridge University Press, Cambridge, 1976.

Gilmore, R.
Lie Groups, Lie Algebras, and Some of their Applications, John Wiley and Sons, New York, 1974.

Golubitsky, M. and Schaeffer, D.
Singularities and Groups in Bifurcation Theory, 1, Springer-Verlag, New York, 1985.

Gottfried, K. and Weisskopf, V.
Concepts of Particle Physics, Oxford University Press, Oxford, 1984.

Hamermesh, M.
Group Theory, Addison-Wesley, Reading, Massachusetts, 1962.

Harish-Chandra.
On some applications of the universal enveloping algebra of a semi-simple Lie algebra, *Trans. Amer. Math Soc.* **70** (1951), 28–96.

Heisenberg, W.
Uber den Bau der Atomkerne, I, II *Z. Physik* **77** (1932) 1–11; **78** (1932), 156–164.

Helgason, S.
Differential Geometry and Symmetric Spaces, Academic Press, New York, 1962; 2nd ed.: *Differential Geometry, Lie Groups, and Symmetric Spaces*, Academic Press, New York, 1978.

Hermann, R.
Lie Groups for Physicists, Benjamin, New York, 1966.

Ince, E.L.
Ordinary Differential Equations, Dover, New York, 1949.

Jacobson, N.
Lie Algebras, Interscience, John Wiley and Sons, New York, 1962.

Jaric, M.V. and Birman, J.L.
New algorithms for the Molien functions, *Jour Math. Phys.* **18** (1977), 1456–1458.

Kaufman, B.
Crystal statistics, II. Partition function evaluated by spinor analysis. *Physical Review* **76** (1949), 1232–1243.

Kirillov, A.A.
Elements of Representation Theory, Springer-Verlag, New York, 1976.

Kostant, B.
The solution of a generalized Toda lattice and representation theory, *Adv. in Math.* **34** (1979), 195–338.

Lax, P.
[1] Integrals of nonlinear equations of evolution and solitary waves, *Comm. of Pure and Applied Math.* **21** (1968), 467–490.
[2] Almost periodic solutions of the KdV equation, *SIAM Review* **18** (1976), 351–375.

Lie, S.
Gesammelte Abhandlungen, bd. 1–7, B.G. Teubner, Leipzig, 1922–1960.

Marsden, J.E.
Lectures on Geometric Methods in Mathematical Physics, CBMS-NSF conference series in applied mathematics, SIAM, Philadelphia, 1981.

Michel, L.
Symmetry defects and broken symmetry. Configuration hidden symmetry. *Rev. Mod Phys.* **52** (1980), 617–651.

Miller, W.
Symmetry Groups and their Applications, Academic Press, New York, 1972.

Montgomery, D. and Zippin, L.
Topological Transformation Groups, Interscience, John Wiley and Sons, New York, 1966.

Olver, P.
Applications of Lie Groups to Differential Equations, Springer-Verlag, New York, 1986.

O'Neill, B.
Elementary Differential Geometry, Academic Press, New York, 1966.

O'Raifeartaigh, L.
Lorenz invariance and internal symmetry. *Phys. Rev. B.* **139** (1965), p. 1052.

Ovsjannikov, L.V.
Group Analysis of Differential Equatibns, Academic Press, New York, 1982.

Pontryagin, L.S.
Topological Groups, 2nd ed., Gordon and Breach, New York, 1966.

Samelson, H.
Notes on Lie Algebras, Van Nostrand, New York, 1969.

Sattinger, D.H.
[1] *Group Theoretic Methods in Bifurcation Theory*, Springer-Verlag Lecture Notes in Mathematics **762**, Springer-Verlag, New York, 1979.
[2] Bifurcation and symmetry breaking in applied mathematics, *Bull Amer Math. Soc.* **3** (1980), pp. 779–819.
[3] *Branching in the Presence of Symmetry*, CBMS-NSF Conference Series in Applied Mathematics, SIAM, Philadelphia, 1983.
[4] Hamiltonian hierarchies on semi-simple Lie algebras, *Studies in Applied Mathematics* **72** (1985), 65–86.

Schultz, T.D., Mattis, D.C. and Lieb, E.H.
Two dimensional Ising model as a soluble problem of many fermions, *Reviews of Modern Physics* **36** (1964), 856–871.

Schwinger, J.
On angular momentum. In: *Quantum Theory of Angular Momentum*, Biedenharn, L.C. and Van Dam, H. (Eds.), Academic Press, New York, 1965.

Singer, I.M. and Thorpe; J.A.,
Lecture Notes on Elementary Topology and Geometry, Springer-Verlag, New York, 1967.

Sternberg, S.
Lectures on Differential Geometry, Prentice-Hall, Englewood Cliffs, New Jersey, 1964.

Struik, D.J.
Classical Differential Geometry, Addison-Wesley, Reading, Massachusetts, 1961.

Symes, W.W.
Systems of Toda type, inverse spectral problems, and representation theory, *Invent. Math.* **59** (1980), 13–51.

Vander bauwhede, A.
Local Bifurcation and Symmetry, Research Notes in Mathematics, **75**, Pitman, 1982.

Varadarajan, V.S.
Lie Groups, Lie Algebras, and Their Representations, Graduate Texts in Mathematics, Springer-Verlag, New York, 1984.

Weyl, H.
[1] Theorie der Darstellung kontinuerlicher halb-einfacher Gruppen durch lineare transformationen, Teilen I, II, III, *Math. Zeit.* **23** (1925), 271–309; **24** (1926), 328–376; **24** (1926), 377–395.

[2] *The Classical Groups*, Princeton University Press, 1946.

[3] *The Theory of Groups and Quantum Mechanics*, Dover, New York.

Whittaker, E.T.
A Treatise on the Analytical Dynamics of Particles and Rigid Bodies, Dover Publications, New York, 1944.

Wigner, E.
[1] On unitary representations of the inhomogeneous Lorenz group, *Ann. Math.* **40** (1939), 149–204.

[2] *Group Theory and Its Application to the Quantum Mechanics of Atomic Spectra*, Academic Press, New York, 1959.

Wybourne, B.G.
Classical Groups for Physicists, John Wiley and Sons, New York, 1974.

Index

Applied Mathematical Sciences

cont. from page ii